Monograph of the Ayrshire Breed of Cattle

The Dairy Cow: With an Appendix on Ayrshire, Jersey and Dutch Cattle Milks

by Lewis Sturtevant and Joseph N. Sturtevant

with an introduction by Jackson Chambers

This work contains material that was originally published in 1875.

This publication is within the Public Domain.

This edition is reprinted for educational purposes
and in accordance with all applicable Federal Laws.

Introduction Copyright 2017 by Jackson Chambers

Self Reliance Books

Get more historic titles on animal and stock breeding, gardening and old fashioned skills by visiting us at:

http://selfreliancebooks.blogspot.com/

Introduction

I am pleased to present another title in the "Cattle" series.

The work is in the Public Domain and is re-printed here in accordance with Federal Laws.

As with all reprinted books of this age that are intended to perfectly reproduce the original edition, considerable pains and effort had to be undertaken to correct fading and sometimes outright damage to existing proofs of this title. At times, this task is quite monumental, requiring an almost total "rebuilding" of some pages from digital proofs of multiple copies. Despite this, imperfections still sometimes exist in the final proof and may detract from the visual appearance of the text.

I hope you enjoy reading this book as much as I enjoyed making it available to readers again.

Jackson Chambers

"Brute foster-mother, mild of humankind,
Whether in farm-yard ruminant reclined
At eve, with richest pasturage distent,
Emblem of rural quiet and content;
From their secretions sweet their udders freed,
Or grazing patiently on hill or mead,
No beast or tame or wild, O gentle cow,
Can sweeter thoughts recall to mind than thou.

"The golden butter is thy produce, and
Thou feedest all the nurseries of the land
With streams nectareous, health-bestowing, sweet,
When iced, a luscious drink in summer's heat;
In the old mythic heaven of the North
The cow Adumbla prominent stood forth.
When summer suns extend their farewell beams,
At eve, what pastoral music sweeter seems
Than the cow's lowings when she hastens home,
While clouds of insects round her sport and hum;
Her breath is then most odorous indeed,
Full of the scent of hillside and of mead;
Inhaling it the milkmaid's cheeks can show
A bloom such as cosmetics can't bestow."

CONTENTS.

I. GENERAL AND DESCRIPTIVE.

	PAGE
THE AYRSHIRE COW IN GENERAL	11
AS A MILKING ANIMAL	24
AS A BUTTER PRODUCER	46
AS A CHEESE PRODUCER	53
MEAT	64
OPINIONS OF THEIR WORTH	69
THEIR ADAPTABILITY	68
THE IDEAL AYRSHIRE COW	70
THE AYRSHIRE BULL	91

II. HISTORY.

SCOTLAND AND ITS PAST	99
WHITE FOREST BREED OF CATTLE	106
COUNTY OF AYRSHIRE	123
DOCUMENTARY HISTORY OF ORIGIN OF AYRSHIRE CATTLE	137
ORIGIN OF AYRSHIRE CATTLE	145
PROGRESS OF THEIR IMPROVEMENT	149

III. LOCAL.

IMPORTERS AND IMPORTATIONS	157
PEDIGREE AND THE HERD BOOK	182
LIST OF IMPORTED PRIZE AYRSHIRES	195
LIST OF WOOD-CUTS OF IMPORTED AYRSHIRES	196
PEDIGREES OF ILLUSTRATED ANIMALS	199

APPENDIX.

MILK: ITS FORMATION AND PECULIARITIES, ETC.	203
AYRSHIRE MILK.	
JERSEY MILK.	
AMERICAN HOLSTEIN MILK.	
CREAM	240

I.

GENERAL AND DESCRIPTIVE.

In this portion of our book we propose to present the Ayrshire cow to the reader in her various appearances and uses.

Our first chapter will tell what the Ayrshire cow is, presenting extraneous matter only as illustrating this feature; in the three following divisions will be given the statements of her products, so far as is known to us; our fifth chapter deals with the testimony of her peculiar merits, while the following one treats the question of adaptability.

In the chapter on the ideal animal which naturally closes our first section, we have aimed to give the Ayrshire cow as she may become, founding our judgment on the teachings of nature and reason, with the attest of careful observation.

AYRSHIRES.

Mr. M'Combie, of Tillyfour, when called upon to read a paper before the Chamber of Agriculture, could begin thus: "My father and grandfather were dealers in cattle." Not ours is this preparation, nor have we "wintered and summered" with the Ayrshire, and heard her nightly breathing below our straw bed in the byre, as happens with many a Scotch hind. It were rare good fortune that should strike out dulness from the mind of one so familiar with his daily care, and unite ideas and the pen in the single hand.

Not the less have we thought of the Ayrshire cow, if we have kept less near to her. How many billows are there, think you, between her home and ours? Answer this, and we will tell you we have braved them all, in order to study her in her home.

Our plan is simply to bring to your ear a narrative of what is known and thought respecting this interesting breed of cattle.

The County of Ayrshire, in the southwest part of Scotland, has given name to a breed of cattle celebrated for their dairy qualities. This county is in outline nearly of the form of a half moon, concave towards the sea and convex on the land side. Ayr, at the joining of river to sea, the most considerable town, is midway between the northern and southern extremities, some eighty miles apart. Although the seat of the origin of the breed, as improved, is placed to the northwestward of Ayr, the cattle have been so long since dispersed over the country, and have been cultivated with such care, that the best may now be found in a region of which no place is much above a dozen miles distant from the home of Burns.

To offer a portrait of this breed, that shall be true to nature, is not so easy a matter as may at first appear. While the individuals possess that in common which clearly portrays their kinship, there is withal much of individuality, as marked by color, form, and quality; but none are so different from their type as to cause a good judge of the stock to think one a member of another breed.

When there is a suggestion of a cross, there yet clings to the Ayrshire an indescribable something, an air, a style, that sets her apart from all others.. Words have not the nice shades of meaning to give conveyance to the thought.

Look at Rosa Bonheur's group, "A Morning in the Highlands." See her brace of shelties resistingly led by the Highland lad. What freshness! The very spirit of nature is there.

The serene, mild expression of the Short-horn comes of breeding; nature, unassisted, not often gives it. It is a subdued, perhaps may become a care-worn look. What a dead look does the ill-bred native cow turn towards us!

The Ayrshire possesses something of the spirit of the English thoroughbred horse. With good treatment, he shows a docile intelligence, ready to perform for you all sorts of kindly offices. No horse can you place such dependence on, none so safe, when well trained, as the thoroughbred; but there is fire enough, enough of nature in him, to outrun a thousand of the cold-blooded kind, and instincts, too, that show that domestication has but regulated and not demeaned him.

With all the high breeding the Ayrshire shows, she is yet near to nature. Breeding, as in the Short-horn, has not made a dull thing of the cow and a harmless thing of the bull. Did you ever see a rabbit in the forest, erect and listening, who has not yet seen your person but has heard your step? There are instincts and nerves here: enough to supply a herd of Short-horns.

The Ayrshire has a superabundance of nerves. She is ready to employ them upon demand, in self-defence or in self-support; she asks little beyond a fair chance: yet all this nature in her is in reserve, and she does not use it wantonly to disqualify her to be the pet of the household. She can the more aptly accommodate herself to circumstances and make them friendly to her.

This wealth of instincts, all alive upon occasion, adapts her to be appreciative of good treatment, and appeals to intelligence to accord it to her.

"If to her share some trifling errors fall,
Look in her face and you'll forget them all."

The head is the seal of character, and bears its stamp. Breeding does much for it. From seeing no other part can one infer so much. If, therefore, you wish to see but one part of the male, study the head above all; in the female, the head and the udder furnish a key to the rest.

Upon this point, Mr. Henry Corbet[1] has written well. "The shoulder, no doubt, answers very much for shape and symmetry and frame, but the head answers for everything. If you go for breed, you look, above all, to the head; if your aim be style or fashion, you must seek this in the head, as nine times in ten that very accommodating phrase known as 'quality' should prove itself by a good head. . . . A scale of points for one or two certain breeds has already been drawn out, but in none of these is sufficient importance, at least as I am led to think, attached to the head of an animal as the main index to his purity of blood, strength of constitution, and actual fitness for that service for which it is intended. . . . Early maturity or quick feeding is the chief recommendation of a Short-horn; and so, when we look one in the face, we must bear in mind that what we want is, as Mr. Carr puts it, 'a placidity and composure of mind, a phlegmatic disposition suggestive of fattening propensity.' In fact, a frisky Short-horn should be something of an anomaly.

"Not so the Devon. I should myself have a fancy for a certain wildness or boldness in the head of a pure North Devon; and when Captain Davy says this

---

[1] Bath and West of Eng. Soc. Jour., quoted in Ohio Ag. Rept. 1871.

should in many points resemble the head of the deer, he seems to me to have very aptly illustrated his subject. Says Captain Davy, 'The head should be small, with a broad, indented forehead, tapering considerably towards the nostrils; the nose of a creamy white; the jaws clean and free from flesh; the eye bright, lively, and prominent, encircled by a deep orange-colored ring; the ears thin, the horns of the cow long, spreading, and gracefully turned up. . . . The expression must be gentle and intelligent. . . . The champion Hereford bull of this day . . . begins with a somewhat mean, small head; whereas there should be something very noble in the head of a White-face, when seen at his best.'

"There is no animal which tells more of high breeding than an Alderney, or rather Jersey-born cow. There is a refined air and carriage, a certain comely 'presence,' which would forbid all thoughts of the butcher, and never carry one's appetite beyond a syllabub on thin bread and butter. Beyond a peculiar, wild, wicked eye, there is not much to admire in the head of an Alderney bull, and even the cows lose much of their graceful character when bred away from their native isle.

"In the Jersey scale of thirty-six points for a perfect cow or heifer, one each is allowed for the following excellences: 'Head small, fine, and tapering; cheek small; throat clean; muzzle fine, and encircled by a light color; nostrils high and open; horns smooth, crumpled, not too thick at base, and tapering; ears small and thin (one point), of a deep orange color within (one point); eye full and placid.' The eye of the bull must be lively, and his horn tipped with black, but beyond these, the points are much the same.

"Mr. M'Combie again, speaking of course of his much beloved black Polls, says: 'A perfect breeding or feeding animal should have a fine expression of countenance; I could point it out, but it is difficult to describe upon paper. It should be mild, serene, and expressive. He should have a small, well-put-on head, prominent eye, with a clean muzzle.'

"Let us," says Mr. Corbet, "look to another kind of Scotch cattle, and what would the West Highlander be without his head? The butcher will say in answer, 'The very best beef'; but with his head all his character is gone. There is a wild grandeur, I had almost said majesty, about the head of the Highlander, that should count up very fast in any scale of his points, as perhaps no other animal shows in this respect such insignia of nature's nobility. You may read of his Highland home in his clear, bright eye, his magnificent horn, and his rough but right royal coat."

The Ayrshire head is not like any of these. Indeed, in these descriptions the most perfect animals have been figured, and not the animal typical of the breed. If many of the Ayrshires hint of the Highland, of which they may inherit something, it is a hint only. Though doubtless something of their unrest and assurance is only half-concealed in her face, there is a cowy or milky look that comes of the use for which she is reserved; there is the look of domestication, but in general, of a domestication that has not been carried to the highest pitch. It has not, as in the Prince Albert Suffolk swine, quite subjected her to its behests. Of course, the degree to which this is carried varies in different families. The countenance should be serene, mild, and expressive, the latter to be born of motherly instincts. The perfect animal is being brought to this, but the majority of the Ayrshires have an earnest liveliness of expression which is all their own, and which the portrait artist must recognize.

In form, the head may be long, and of no great comparative breadth, or it may be short, with considerable breadth.

The short head has come from such breeding as Theophilus Parton, of Swinley farm, pursued, and it is known as Swinley stock. This stock differs from the older stock in having a shorter head, with more breadth across the eyes, more upright and spreading horn, more hair, and that of a more mossy character, and generally better constitution.[2]

The points for the head, given by the Ayrshire Agricultural Association in 1853[3] as indicating superior quality, are as follows: "Head short, forehead wide; nose fine between the muzzle and the eyes; muzzle moderately large; eyes full and lively; horns widely set on, inclining upwards, and curving slightly inwards."

---

[2] Sandford Howard's article in W. S. Dept. Ag. Rept. 1863, p. 195.
[3] Prize Essays High. and Ag. Soc. 1866-7, p. 106.

William Aiton, in the survey of Ayrshire, printed at Glasgow in 1811, says the shapes most approved of are, "Head small, but rather long and narrow at the muzzle. The eyes small, but smart and lively. The horns small, clear, crooked, and their roots at considerable distance from each other."

These aspects, and a compromise of them in varying degree, are found in the Ayrshire of to-day.

The carriage is what may be inferred from a study of the head of the animal. Each motion is suggested by a purpose entertained by her, and her walk is seldom lagging; and if she pauses by the way-side, it is but for a moment, to move on at a quicker pace. There is little dilatoriness. Promptness is a characteristic. Her walk is easy, hurried into a trot in the early morning, and at night, if she expects to find food in her manger, or to drink there. If you disturb her at rest, in the pasture, she goes to feeding again.

There is often too much motion for her to be graceful. She steps precisely and long, but when grazing, no animal can be more pleasing. Her shapes are so carried as to offer small impediment to motion, and it comes easier to her than to any other dairy breed in our acquaintance that carries so much of the pasture with them.

In the dairy breeds, and in most animals particularly adapted to milk-giving, there is a tendency towards accumulation of a larger part of the weight of the animal in the rearmost half. In the Ayrshire, this tendency is much developed, more so than in any other breed whatsoever. As judged by a side view, or from above, there is a certain wedge form. Although in this breed the shoulders lie close, this wedge shape is derived less from a deficiency forward than from the large bulk of the carcass aft. This form becomes more strongly marked with age, when the animal has been abundantly supplied with food. The yearling and two years' old may have parallel rather than diverging lines on the side view.

By referring to several descriptions or "scales of points" to which it has at various times been judged that this breed should conform, we may derive a tolerably clear idea of its present appearance.

It must, however, be borne in mind that the possession of these points by an animal is exceptional rather than common, but the study of them directs us to what is typical of the breed. They are made up not from diverse breeds, nor are they ideal, but have existed either in conjunction in some exceptionally fine animal of the breed, or have been observed separately.

	1853.[4]	1829.[5]	1811.[6]
Head.	Short, forehead wide.	Small, long and narrow towards muzzle.	Small, but rather long and narrow at the muzzle.
Nose.	Fine bet. the muzzle and eyes.		
Muzzle.	Moderately large.		
Eyes.	Full and lively.	Not large, but brisk and lively.	Small, but smart and lively.
Horns.	Wide set on, inclining upwards, and curving slightly inwards.	Small, clear, bent, and placed at a considerable distance from each other.	Small, clear, crooked, and their roots at considerable distance from each other.

[4] Prize Essays High and Ag. Soc. 1866–7, p. 106.
[5] William Harley, Harleian Diary System, p. 106.
[6] Aiton, Survey of Ayrshire. Glasgow, 1811. p. 426.

Sequence changed from the authorities, but substance given with exactness.

SCALE OF POINTS.

Neck.	Long and straight from the head to the tip of the shoulder; free from loose skin on the under side, fine at its junction with the head, and the muscles symmetrically enlarging towards the shoulders.	Slender and long, tapering towards the head, and a little loose skin below.	Long and slender, tapering towards the head, with no loose skin below.
Shoulders.	Thin at the top.	Thin.	Thin.
Brisket.	Light.		
Forequarters	The whole forequarters thin in front, and gradually increasing in depth and thickness backward.	Light	Light.
Hindquarters.		Large and broad.	Large.
Back.	Short and straight.	Straight.	Straight, broad behind.
Spine.	Well defined, especially at shoulders.		
Joints of Spine.		Slack and open.	Rather loose and open.
Short Ribs.	Arched.		
Body.	Deep at the flanks.	Carcass deep in the rib.	Carcass deep.
Pelvis.	Long, broad, and straight.		Capacious and wide over hips.
Buttocks.			Round and fleshy.
Hook Loins.	Wide apart, and not much overlaid with fat.		
Thighs.	Deep and broad.		
Tail.	Long and slender, and set on level with the back.	Small and long, reaching to the heels.	Long and small.
Legs.	Short, the bones fine, and the joints firm.	Small and short, with firm joints.	Small and short, with firm joints.
Milk Vessel.	Capacious, and extending well forward, hinder part broad, and firmly attached to the body; the sole or under surface nearly level.	Capacious, stretching forward, square, but a little oblong, not low hung, thin skin'd.	Capacious, stretching forward, broad and square, neither fleshy, low hung, nor loose.
Teats.	From two to two and a half inches in length, equal in thickness, and hanging perpendicularly; their distance apart at the sides should be equal to about one third of the length of the vessel, and across to about one half of the breadth.	Small, pointing outward, and at a considerable distance from each other.	Short, all pointing outward, and at a considerable distance from each other.

Milk Veins.	Well developed.	Capacious and prominent.	Large and prominent.
Skin.	Soft and elastic.	Loose, thin, and soft like a glove.	Thin and loose.
Hair.	Soft, close, woolly.	Short, soft, and woolly.	Soft and woolly.
Figure.		Handsome and well proportioned.	Compact and well proportioned.
Temper.		Quiet and docile.	
Color.	Preferred brown, or brown and white, the colors being distinctly defined.		

In nothing does the Ayrshire cow show breeding more than in the milk-vessel or udder. Nowhere, we are tempted to say, can the art of breeding show a greater triumph. Not that all Ayrshires have perfection of form in udder, yet very many approach it. The more skilfully bred indicate the fact here more broadly than in any other particular. We find here oftentimes the stamp of the insignia of art when there is much of naturalness in the surrounding parts. Here is found the index by which the breeder can, in a measure, gauge the degree of removal from the primitive state.

The udder has been the point towards which the search after quality has been directed by the careful Scotchman for a long period of time. Although it differs in outward shape in individuals, it yet retains a certain uniformity which may be considered typical. This is in the gland and the teat. The glands are rather flattened, than pointed as in the Alderney, or elongated as in the Dutch. These are well held up to the body, and, in the types of the breed, extend far forward and back, with a broad and level sole. The teats are small, and of a cylindrical shape rather than cone-shaped, as seen in the Alderney and other breeds. This udder is admirably fitted, by its elasticity, for the storage of milk, and when the glands are at rest, occupies but a small space. The eye, accustomed to seeing the pendent fleshy udder so often met with in dairy cows, is apt to underrate, in comparison, the capacity of the small bag of this breed, with its wrinkled and folded covering, so deceptive to the unskilled, so full of promise of deeds of worth to the educated observer. Fill out these wrinkles and expand these folds, and the lusty calf may well forget his greed at the sight of the stores at his disposal.

The Scotch having been less intent to secure a particular color than quality in their herds, although exercising some taste in the matter, their cattle, as do those of the Channel Isles, offer much variety to the eye. There are among them no such mixtures as red and white, so mixed as to be a roan, or black and white thus disposed. It is rare for one color to mingle with another; the line of separation being generally distinct.

Of 236 animals imported into the United States, about 70 per cent are described as red, or red and white. Of 2,852 animals in the United States whose colors are given, about 78 per cent are called red, or red and white.

The following table,[7] although the descriptions of color may not be strictly accurate, as there is probably little preciseness in recording shades, may be of interest:—

---

[7] Am. and Can. A. H. B.

Described as	Number of Animals.	Percentage
Red, or mostly red	222	7.78
Red and white	2,014	70.61
Brown, or mostly brown	47	1.64
Brown and white	241	8.45
Mostly black	2	?
Black and white	3	?
Yellow	1	?
Yellow and white	24	.84
Dun	4	?
Brindle	20	.70
White, or mostly white	17	.59
White and red	194	6.80
White and brown	19	.66
Fawn	2	?
Roan	2	?
Spotted, flecked, etc. etc.	40	1.40

Were all the animals here recorded known to be, without question, of the pure breed, the occurrence of the few anomalous colors would open a discussion of great interest. Of the imported animals, in but few do anomalous colors occur; one is described as bay and white, and two as brindled.

We have never seen one all white; to find one seven eighths white, with red or brown ears and cheeks, is not very rare. Black, or black and white, occur, but are not common. Some may be brindled in part, as black and brown mixed. Although this is the result of our own obsevations in Ayrshire, yet it may be well to quote from a letter of Robert Wilson, a most intelligent man and successful breeder. "Colors of Ayrshires are much the same since I can remember; different breeders have their particular color. Light yellow, though common with some breeders, is not the most common color. Red and white flecked, though it should incline a little to yellowish or brown, is more a prevailing color of the breed. . . . White, if there be not roan mixed with it, I do not consider a proof of the presence of short-horn blood. Our favorite colors are white flecked, or red bodies and white legs. Dark reds and black muzzles are favorites also. This color is considered the hardiest, though I incline to think there is somewhat of a cross in it. Ayrshires are not disqualified as prize-takers on account of color."

In Ayrshire, the design all along has been to discourage the growth of those points which, though perhaps useful in the animal reared for meat, have no use in the dairy animal. That she yields much milk, and that she yields it without extravagance of food, is the end sought. Everything in the economy of the animal must be subsidiary to this; and if any one can point out in her figure a pound of flesh that is not tributary to this purpose, or if there is needless weight of bone, then it will fall to the breeder to lop it off. It is a characteristic of the Ayrshire that she carries her weight only, and lives only, to serve dairy interests with the utmost utilization of food.

But her service in this direction does not preclude her from taking on flesh rapidly when not in milk, and fed well, nor does it preclude the meat being of the best quality. Though she does not afford the butcher meat in as economically-shaped pieces as the Short-horn, so like a brick, in form of carcass, it is equally as good, if not superior. In the Ayrshire, the fat is mixed with the lean, evenly and in thin streaks. When fed for the butcher, then all her energies are directed to meet his demands, the food that has hitherto gone to milk being directed to an equalization of flesh over the whole animal. The aged beast thus fattens readily and economically, and furnishes a flesh of a juicy texture and high quality.

The Ayrshire cow is a renowned milker through inheritance; yet the Scotch have a saying, taught by experience, that "the cow gives her milk by the mou'." It is a fancy of the sculptor that the figure he is about to cut already exists in the marble, and it is his work only to expose it to view. So may we, employing our fancy, see milk lying concealed in the grasses, which the cow has but to lap in order to fill the pail.

The food, and the machine for the conversion of food into milk, are the two elements that, united in a happy manner in one harmonious design, make the production of milk a commonplace affair. But who shall raise it from the commonplace by exposing the secret springs of action, and prying into the concealments of nature?

The question of milk, however, deserves a chapter of itself, where it can be treated in a manner commensurate with its importance.

THE AYRSHIRE AS A MILKING ANIMAL.

That the Ayrshire cow is a large milker there can be little doubt, as the fact is supported by universal testimony. Yet it may be well to present those statements of yields which we have collected.

Ro. Forsyth,[1] writing before 1807, says that "twelve of these small cows will yield for four or five months in succession 120 Scotch pints of milk each day." As the Scotch pint is $103\frac{4}{10}$ cubic inches, this would be nearly 18 quarts per cow.

Aiton,[2] writing in 1811, says that some of the dairy cows in Ayrshire may yield for a time from 12 to 14 Scotch pints ($21\frac{1}{2}$ to 25 wine quarts) per day, but such returns are rare. Many of them will, when in their best plight, and duly fed, yield at the rate of 10 Scotch pints (about 18 quarts) of milk per day for two or three months, probably about 6 pints ($10\frac{3}{4}$ quarts) for other three months, and say 3 pints ($5\frac{1}{3}$ quarts) for four months more, making in all during the season about 1,700 or 1,800 pints (3,046 to 3,225 quarts). Many cows, however, will not yield more than the half of that quantity. Probably 1,200 Scotch pints (2,148 quarts) of milk from each cow in the course of the year, may be about a fair average

---

[1] Beauties of Scotland, iii, 8.   [2] Aiton's Survey of Ayrshire, p. 464.

MILK YIELD.

of the Ayrshire dairy stock. He had heard of 16 or 18 pints (28½ to 32 quarts) being taken from a cow every day, but had never seen so much.[3]

In 1829 William Harley states "as the average for the Harleian dairy, 12 quarts per day." This is 4,380 quarts a year. It will be remembered, however, that in this dairy the cattle were kept in very high condition, and were continually being turned for the butcher; and these high figures do not represent the average for a cow, but only for the average number kept during the year. Mr. Harley bought one very large fine cow at a high price. This cow gave for a considerable time 40 quarts a day. He had a number of other very fine cows which, when newly calved and highly fed, produced from 25 to 30 quarts per day.[4]

Dr. Voelcker,[5] of England, writing in 1863, mentions a cow bought by the Duke of Athol from Mr. Wallace, Kirklandholm, and probably in his Grace's dairy at Dunkeld House at the present time, that produced 13,456 pounds (6,258 quarts) of milk from the 11th of April, 1860, to the 11th of April, 1861.[6]

In Morton's Farmer's Almanack for 1866, the average annual yield per cow in five known dairies is given at 4,992 pints, but which is stated to be above the average of ordinary grass-fed cows.[7]

Aiton's Survey of Ayrshire, p. 428.
Harleian Dairy System, pp 87 and 106.
Jour. of R. A S. of Eng. 1863, p. 308.
Journ. R. A. S. of Eng. 1863, p. 308.
Quoted in Pr. Essays H. Soc. 1866-7, p. 78

MILK YIELD.

In Mr. Buttery's dairy of thirty cows, at Calder Bank, the average annual produce is 2,941 wine quarts per cow.[8]

At the competition between milch cows at the show of the Ayrshire Agricultural Society, the average milk yield for four milkings was 49^1 pounds a day, as follows[9]: —

Owner.	Weight of 4 Milkings during 2 days.	Per day.	Per milking.
Archibald Wilson....	96 lbs. 14 oz.	48 lbs. 7 oz.	24 lbs. 3¼ oz.
James Hendrie.......	97 " 4 "	48 " 10 "	24 " 5 "
William Reid........	82 " 3 "	41 " 1½ "	20 " 9 "
do.	109 " 6 "	54 " 11 "	27 " 5¼ "
R. Wallace..........	114 " 2 "	57 " 11¼ "	29 " 8¼ "
do.	94 " 1 "	47 " ½ "	23 " 8¼ "

Archibald Sturrock, in 1866, estimates the average yield for all the cows at about 3,400 imperial pints from each cow per annum, and apologizes for his low estimate by complaining of the want of house feeding by so many, and the great neglect of the cows in winter.[10] 3,400 imperial pints is 2,040 wine quarts.

To complete the records of yield in Scotland, we will quote from statements made us, either in person or by letter, from Scotch breeders.

Robert Wilson, of Kilbarchan, writes, "I have known cows in our own stock to give as much as 28 and 32 quarts each, daily, but such are exceptional cases."

At Kilmarnock, in 1869, we found the universal testimony was from 12 to 16 quarts daily, from the best cows.

---

[8] Mayne's How to Choose a Milch Cow, p. 136.
[9] Gard. Chron. and Ag. Gazette, Apr. 8, 1865.
[10] Prize Essays H. Soc. 1866-7, p. 78.

Mr. Ormsby, near Ayr, states the average yield of good cows, for three months in succession, as from 7 to 8 pints at a milking; that is, 14 to 16 quarts daily.

Mr. Robert McKeen, of Bishopbriggs, near Glasgow, had, in 1869, a herd of 36 very superior-looking animals. He gave their yield as $21\frac{1}{2}$ quarts for three months. He was a high feeder, and his nearness to the breweries of Glasgow gave him the privilege of obtaining brewery draff, which, it is needless to say, was abundantly availed of.

Professor Wilson, in his report on the Danish Exhibition, says, "Two dairies of Ayrshire cows gave the annual milk product per cow at 2,600 quarts and 2,528 quarts respectively."[11]

In America we can hardly expect as large an annual return in milk as obtains in Scotland, on account of the greater dryness of our climate, and the absence of that succulency of food, during the summer heats, which is so desirable. Yet on account of the care exercised towards cows so highly valued, we have instances of excellent yields for the year, and in the yields for a limited period oftentimes statements so remarkable for their excess as to call for further verification. We, however, give our authorities in each instance.

Of the four cows imported by Mr. Cushing in 1837, we have the following memoranda of their yields commencing in that year.[12]

---

[11] Trans. N. Y. Ag. Soc. 1869, p. 666.
[12] Farmers' Library, iii, 305.

FLORA.

From May 17 to June 1..	608 lbs.
In June	1,192 "
In July	1,064 "
In August	841 "
In September	718 "
In October	489 "
In November	409 "
In December	432 "
In January, 1838	442 "
In February	388 "
In March	484 "
In April	419 "
20 days in May	242 "
Total	7,728 "

VENUS.

From June 20 to July 1.	283 lbs.
In July	805 "
In August	693 "
In September	567 "
In October	498 "
In November	319 "
In December	403 "
In January, 1838	406 "
In February	351 "
In March	368 "
In April	319 "
21 days in May	151 "
Total	5,163 "

JUNO.

From May 23 to June 1..	243 lbs.
In June	796 "
In July	845 "
In August	600 "
In September	475 "
In October	313 "
In November	340 "
In December	394 "
In January, 1838	401 "
In February	326 "
In March	328 "
In April	216 "
7 days in May	30 "
Total	5,307 "

CORA.

From Nov. 17 to Dec. 1..	388 lbs.
In December	834 "
In January, 1838	846 "
In February	776 "
In March	704 "
In April	670 "
21 days in May	405 "
Total	4,623 "

Mr. F. H. Appleton, of West Peabody, Mass., gives the yield of three of his cows for the year commencing August 26, 1871, and ending August 25, 1872, at 8,159¾ lbs., 7,728¾ lbs., and 5,277¼ lbs. respectively.[13]

The farm year of Mr. E. T. Miles, of Fitchburg, Mass., commences on the 1st of July, and we transcribe the records of Maplewood Farm in full.[14]

---

[13] Trans. of Essex Co. Ag. Soc. 1872, p. 74.
[14] Milk Record of Maplewood Herd (Ayrshire), Fitchburg, July 1, 1872, also, do. 1873. Also, MS. communication from Mr. Miles.

	Age July 1, 1872.	Weight, 1872.	No. days in milk.				Yield of milk in lbs.			
			'69 & '70.	'70 & '71.	'71 & '72.	'72 & '73.	'69 & '70.	'70 & '71.	'71 & '72.	'72 & '73.
Miller, 2d	13	1,110	284	279	288	282	6,588½	4,797	5,692¾	5,764¼
Beauty	11	985	315	365	298	317	8,011	7.922½	7,555¼	7,304¾
Emma	11	1,070	284	307	280	247	5,831	5,930	4,248¼	4,469¾
Daisy	11	1,125	321	278	263	327	6,953	5,951	5,611¾	6,095¼
Daisy	10	1,028	313	304	302	308	6,618	6,195½	6,300¾	6,526¼
Myrtle, 1st . . .	5	995	265	267	288	297	4,819	5,950½	7,047	7,267
Cleopatra	5	1,085	301	309	324	294	5,178	6,021½	5,764¼	6,982¼
Maud Muller . .	5	1,200	. .	. .	295	319	. . .	. . .	5,493	5,880¾
Lady Burns . . .	5	1,073	163	307	284	292	3,190	5,102½	4,785¼	5,331¾
Ellen Douglas . .	4	848	. .	166	267	253	. . .	3,281	5,313¼	4,916¼
Myrtle, 2d	3	975	. .	. .	114	280	. . .	. . .	2,352¾	5,806
Lady Sampson . .	2	800	. .	. .	. .	271	. . .	. . .	. . .	4,679¾
Lady Burns, 1st .	2	795	. .	. .	. .	267	. . .	. . .	. . .	4,559
Vallonia	2	715	. .	. .	. .	273	. . .	. . .	. . .	4,146¼
Gracie	2	870	. .	. .	. .	229	. . .	. . .	. . .	3,365¼

SUMMARY FOR THE ENTIRE HERD.

Year.	No. of cows.	Av. milk season.	Milk per cow in lbs.	In qts.
1869 and '70	7½	300 days.	6,292 lbs.	2,926 qts.
1870 " '71	8½	303 "	6,017 "	2,798 "
1871 " '72	10½	286 "	5,730 "	2,665 "
1872 " '73	15	283 "	5,539 "	2,530 "
Average	10¼ cows.	293 days.	5,821 lbs.	2,707 qts.

After the above was in press, we received the yield of this herd for the year 1873–4 : —

	Days in Milk.	Yield, lbs.	Quarts.
Miller, 2d	256	4,848¾	2,255
Beauty	322	7,857½	3,655
Emma	290	6,109	2,841
Daisy	300	7,086¼	3,296
Daisy	301	7,358¼	3,422
Myrtle, 1st	323	7,702¼	3,582
Cleopatra	259	5,653¼	2,629
Maud Muller	215	3,622¼	1,685
Lady Burns	300	5,882¼	2,736
Ellen Douglas			
Myrtle, 2d	238	4,344¾	2,021
Lady Sampson	251	3,926	1,826
Lady Burns, 1st	284	4,384¾	2,039
Vallonia	291	3,216¼	1,496
Gracie	295	4,660¼	2,167
14 Cows . . . . . . Average	280	5,475	2,500

Average per cow for five years, 2,642 quarts.

We have now given all the annual yields in our possession, excepting those of the Ayrshires of Waushakum Herd, for which records, on account of their completeness, we reserve a separate paragraph. We will now proceed to give the records of milkings for a period of time less than the year's yield.

First in order comes Ayrshire Lass and Red Rose, whose owner, Mr. James Brodie, presents at the New York Fair, and takes oath to the record whose summary we give.[15]

Ayrshire Lass. 11 years old; calved April 1, 1861. June 10 gave 74 pounds of milk. Commenced August with 66 pounds, and closed with 55 pounds; total for the month, 1,902 pounds. September 1, $55\frac{1}{2}$ pounds; September 16, 51 pounds; first sixteen days of September, 844 pounds.

Red Rose. 8 years old; calved May 20, 1861. June 10 gave 84 pounds of milk. Commenced August with 74 pounds and closed with 63 pounds. Total for the month 2,168 pounds. September 1, 62 pounds; September 14, 50 pounds; first fourteen days of September, $788\frac{1}{2}$ pounds.

Mr. H. H. Peters, in his catalogue for 1865, gives the yield of his cows, Corslet and Jean Armour, as follows: —

Corslet averaged from May 2 to September 1, $21\frac{5}{8}$ quarts of milk per day. The greatest yield was 26 quarts.

Jean Armour, in 1862, gave an average of 49 pounds 3 ounces of milk per day for 114 days, commencing June 1. Greatest yield, 58 pounds a day;

---

[15] Trans. New York Ag. Soc. 1861, p. 125.

least yield, 43 pounds. For the month of July she averaged 51 pounds 13 ounces per day.

One of our neighbors, Mr. Isaac Felch, allows us to take from his book the following record of the yield of his cow Mary, eight years old. She dropped a calf December 2, 1870, and was purchased by him April 19, 1871.

Week ending May 6, 1871	. . 98 qts.	Week ending July 8, 1871	. . 112 qts.
" " " 14, "	. . 100 "	" " " 15, "	. . 105 "
" " " 21, "	. . 105 "	" " " 22, "	. . 105 "
" " " 28, "	. . 119 "	" " " 29, "	. . 105 "
" " June 3, "	. . 123 "	" " Aug. 6, "	. . 100 "
" " " 10, "	. . 126 "	" " " 13, "	. . 98 "
" " " 17, "	. . 123 "	" " " 20, "	. . 70 "
" " " 24, "	. . 119 "	" " " 27, "	. . 63 "
" " July 1, "	. . 119 "	" " " 30, 3 d'ys	. . 31 "

Total, 122 days, commencing 149 days from calving, 1,821 quarts.

Mr. Felch milks himself, and records the measure in his note-book at the time. The cow was in a very fat condition, as Mr. Felch is not only an extremely liberal feeder, but a very careful one.

Mr. Charles Shepherd, of Ogdensburg, N. Y., writes us that he weighed the milk of several of his cows in July, 1869, and it would run from 42 to 50 pounds daily.

Messrs. S. M. & D. Wells, of Wethersfield, Conn., writes that the yield of one of their cows in April is 54 pounds per day. Last week, 50 pounds; week before, 49 pounds.

Mr. J. C. Rutherford, of Waddington, N. Y., writes that in 1870 the average of milk per cow from May 1 to October 1, five months, was $38\frac{1}{2}$ pounds, on grass alone.

Mr. B. Harrington, of Worcester, Mass., writes us as the average of seven cows, 18$\tfrac{3}{7}$ quarts a day, and 16$\tfrac{5}{7}$ quarts for four months.

Mr. Luke Sweetser, of Amherst, Mass., writes us that one of his cows, weighing but 860 pounds, six years old, gave 300 pounds of milk in seven days, and that his cows have ranged from 30 to 50 pounds a day.

Mr. J. D. W. French, of North Andover, Mass., writes us that his cow Dolly gave 2,471 pounds of milk from June 18, 1871, to Sept. 4, 1871, when the record was interrupted by the sending of the cow to the Fairs.

Mr. A. P. Ball, of Stanstead, P. Q., writes us that his yield is 16 quarts in summer and 8 quarts per cow now (October).

Mr. Thomas Miller, of Delaware Co., N. Y., writes us that one of his cows, thirteen days from calving, was giving 55 pounds daily, while another in June averaged 24 quarts.

Mr. J. C. Converse, of Jefferson Co., N. Y., writes that one of his cows, as a two-year old, gave 40 pounds daily, and as a cow, was giving 55 pounds daily in June, and in July, 1871, 45 pounds on pasture.

In 1873 Gen. S. D. Hungerford, of Adams, Jefferson County, N. Y., exhibited at the New York State Fair at Albany an Ayrshire[16] cow known as Old Creamer, whose yield of milk has never to our knowledge been surpassed.

---

[16] This account taken from a card appended to her photograph, sent us by General H. She is probably seven eighths Ayrshire.

CHAMPION COW.

Old Creamer is nine years old, and weighs 1,080 pounds. In three days she yielded the enormous quantity of 302 pounds of milk, as follows: June 11, 100½ pounds; June 12, 100 pounds; June 13, 101½ pounds. She gave 2,820 pounds of milk in the month of June, an average of over 94 pounds per day; 2,483½ pounds in the month of July, an average of over 80 pounds per day.

This cow attracted so much attention at the Fair from the statements of Gen. Hungerford, her admirable form, and evident great capacity for milk-giving, that we annex the following measurements, which we believe will prove of value and interest.

These measurements were taken at noon, September 25, 1873, when she was receiving pails of "slop" every few hours, and was said to be milking 25 pounds three times daily:—

Length of head	19¼	inches.
Breadth between eyes	8	"
Distance around muzzle	18	"
From base of horn to shoulder	27¼	"
From shoulder along spine, to hip	34¼	"
From hip to tail insertion	10	"
Hip points, apart	20	"
Hip point to hook bone	20	"
Depth of flank	22¼	"
Girth	71¼	"
Girth about belly, largest part	96¼	"

Udder oval, broad, extending very far back. Skin loose upwards, and hanging in folds from the vulva.

Milk veins large, equal on either side.

Length of udder	18	"
Depth of udder, gland portion	20	"
Distance along gland from front to rear	29¼	"
Circumference	58	"

Escutcheon 4¼ inches broad just beneath vulva,
   and correspondingly extensive.
Mirrors large.
Disposition very quiet.
Joints of vertebræ loose and open.
Skin medium thick, soft, and easily lifted.

The milk record of Waushakum Farm has now been kept for a number of years under the same system, and we present it with the more confidence as we are personally cognizant of the general correctness of the facts set forth. Commencing with the most carefully-selected native cattle, so called, but for the most part unknown grades, we gradually worked into an Ayrshire-breeding herd, as our trials thoroughly convinced us of their worth. In the following tables are given the results for each cow for each month of her milking, considering only those animals which were kept throughout the farm year, and counting the heifers as cows from the time they came into milk.

WAUSHAKUM MILK RECORD.

WEIGHT OF MILK EACH MONTH, IN POUNDS.
1867.

NAMES OF COWS.	Jan.	Feb.	Mar.	Apr.	May.	June.	July.	Aug.	Sept.	Oct.	Nov.	Dec.	Total in Year.	No. days dry.
Wells	145	1	...	...	939	944	933	858	685	620	445	406	5,976	88
1 Durfee	219	47	2	...	631	1,001	976	927	772	707	519	615	6,416	67
1 Leadbetter	239	47	3	809	914	859	808	785	601	593	497	522	6,677	40
Woodman	226	131	59	132	966	854	675	608	510	495	351	356	5,363	26
2 Leadbetter	204	95	10	153	896	857	841	762	576	561	468	490	5,913	46
Widow	197	51	...	9	843	855	822	753	561	492	461	480	5,524	88
1 Carver	253	136	52	...	425	803	717	687	600	616	557	542	5,388	42
3 Leadbetter	36	...	36	580	650	571	544	488	373	362	286	212	4,138	69
1 Thompson	25	...	150	728	870	847	769	717	517	411	279	302	5,615	66
Carson	385	336	396	403	538	463	306	28	...	654	626	641	4,776	59
Smith	373	357	385	353	11	...	...	...	...	422	1,032	864	3,796	168
2 Carver	152	13	...	717	208	934	898	808	684	661	563	487	5,408	98
Underwood	219	43	21	114	1,014	928	794	715	532	519	233	...	5,735	86
2 Durfee	95	...	...	355	936	792	280	378	327	339	318	401	3,980	83
No. 2	342	221	298	398	414	421	406	84	303	848	212	...	3,906	51
No. 3	464	382	423	373	21	...	...	...	...	...	...	...	1,688	239
No. 4	348	327	349	397	21	...	...	...	...	...	...	317	1,735	226
No. 5	480	403	407	474	21	...	...	...	34	...	...	...	1,708	241
No. 6	609	508	573	522	587	561	389	300	428	652	630	549	5,866	19
No. 8	556	486	511	527	646	662	640	543	454	468	377	433	6,272	...
No. 9	515	469	499	713	676	630	649	539	514	473	277	...	5,738	39
A	...	537	787	345	862	780	664	600	...	510	372	416	6,755	37
B	156	326	363	...	20	...	...	...	...	...	...	570	1,780	214
C	229	445	477	481	579	613	612	491	274	42	...	...	4,243	78
D	245	466	485	449	545	532	484	405	240	169	...	456	4,476	32
Total 25													4,675	87

N. B. — All these cows so-called natives, yet really grades.

WAUSHAKUM MILK RECORD.

1868.

NAMES OF COWS.	Jan.	Feb.	Mar.	April.	May.	June.	July.	Aug.	Sept.	Oct.	Nov.	Dec.	Total in Year.	No. days dry.
Wells............	341	341	267	...	...	...	...	365	598	473	499	329	3,213	141
1 Durfee.........	588	514	590	541	803	504	457	364	193	...	...	...	4,554	96
1 Leadbetter.....	446	402	499	405	502	369	335	284	281	379	406	368	4,676	...
Woodman.........	345	315	383	362	356	...	...	...	...	6	754	765	3,286	172
2 Leadbetter.....	448	429	390	282	74	365	274	483	693	635	670	665	4,789	88
Widow............	465	445	503	454	413	382	368	65	361	392	447	418	4,601	21
1 Carver.........	488	435	499	403	358	468	824	369	298	337	346	278	4,561	...
3 Leadbetter.....	124	20	...	806	774	692	598	706	486	409	389	347	3,693	123
1 Thompson.......	129	241	1,058	556	504	493	448	512	283	276	279	173	5,821	24
Carson...........	608	600	655	618	553	502	441	382	275	297	292	248	5,358	...
Smith............	800	752	802	385	131	...	...	311	67	...	...	...	4,846	101
2 Carver.........	455	519	494	263	18	761	784	622	475	561	524	530	4,079	105
Underwood........	...	...	35	330	320	341	332	331	430	409	446	424	3,567	118
2 Durfee.........	375	360	383	791	748	734	539	539	196	235	246	118	7,055	...
No. 4............	861	849	966	437	383	119	622	...	367	299	273	106	4,437	...
No. 6............	506	442	499	266	330	427	403	358	...	506	830	735	3,824	122
A................	398	369	419	795	776	745	622	580	244	246	225	139	8,037	...
B................	875	850	879	968	847	655	581	547	486	466	506	461	7,690	...
C................	907	1,119	1,113	421	392	390	356	297	358	304	209	84	3,762	...
D................	445	363	477	342	173	12	795	715	204	189	164	64	5,358	30
Cushing..........	488	431	462	431	436	543	484	449	539	479	477	445	2,377	215
Morse............	491	490	529	317	572	701	546	385	347	246	...	...	4,449	34
Lucy.............	526	446	375	717	675	...	...	576	169	295	85	353	5,712	73
Seaverns.........	721	659	784	257	217	...	...	621	466	2	...	371	3,688	61
Fleetfoot 2d, Jersey.	351	366	365	27	...	72	...	23	578	394	389	592	3,671	114
Rose, Ayrshire...	290	277	186	493	377	...	840	724	766	556	544	679	5,567	76
Jennie, Ayrshire.	565	580	602	590	1,071	1,062	559	450	608	728	680	635	6,965	78
Tilly, Ayrshire..	105	...	...	714	683	694	...	...	163	679	671	285	6,334	68
Fanny, Ayrshire..	1,090	874	822											
Total........29													Average........	4,834	64

N. B.—One Jersey, four Ayrshires; the rest so-called natives, really grades.

WAUSHAKUM MILK RECORD.

1869.
WEIGHT OF MILK EACH MONTH, IN POUNDS.

NAMES OF COWS.	Jan.	Feb.	Mar.	April.	May.	June.	July.	Aug.	Sept.	Oct.	Nov.	Dec.	Total in Year.	No. days dry.
1 Durfee	317	701	755	672	720	661	630	540	524	506	254	151	6,431	...
Widow	421	332	396	339	411	315	237	215	409	379	347	339	4,140	...
Smith	252	41	...	406	650	620	609	433	101	...	...	...	3,112	170
2 Carver	482	380	414	424	433	456	445	361	236	145	14	393	3,790	53
Underwood	333	145	20	189	895	746	647	493	348	245	246	393	4,750	45
No. 4	...	343	894	852	722	663	589	440	312	205	115	9	5,144	72
No. 6	645	460	470	449	443	408	363	195	37	...	423	684	4,577	60
B	417	310	358	364	334	198	...	...	...	208	706	706	3,601	115
Cushing	415	301	334	280	120	...	122	706	661	570	418	373	4,300	64
Farm	388	252	217	121	5	608	...	96	547	468	400	533	3,227	115
Grafton	357	750	752	672	641	608	552	517	608	467	370	235	6,529	...
Fay	573	423	484	449	486	470	371	192	14	...	425	280	3,462	105
Vermont	417	253	203	15	...	134	837	693	693	567	425	280	4,517	76
Winthrop	467	315	194	13	77	1,000	900	836	750	541	440	395	5,928	54
Fanny, Jersey	...	...	...	...	259	217	197	289	246	176	71	...	1,449	162
Fleetfoot 2d, Jersey	324	173	41	...	...	15	555	493	522	417	254	241	3,040	105
Rose, Ayrshire	456	228	147	3	...	...	488	594	572	430	284	272	3,474	95
Jennie, "	625	496	525	480	429	261	29	...	367	301	232	426	4,171	53
Tilly, "	571	468	485	448	422	365	312	278	257	163	...	...	3,769	61
Fanny, "	498	513	507	457	470	457	446	356	248	94	1	381	4,378	58
Total........ 20													4,179	73

Average..........

N. B.—Two Jerseys, four Ayrshires; rest so-called natives.

WAUSHAKUM MILK RECORD.

1870.

NAMES OF COWS.	Jan.	Feb.	Mar.	Apr.	May.	June.	July.	Aug.	Sept.	Oct.	Nov.	Dec.	Total in Year.	No. days dry.
No. 4	...	617	989	827	798	648	579	559	489	379	378	340	6,603	38
No. 6	566	546	718	655	617	521	498	489	429	343	69	77	5,528	16
Farn	510	462	572	521	481	345	386	376	345	296	41	...	4,385	49
Fay	497	967	1,252	1,031	1,000	734	617	599	538	258	3	...	7,496	74
Winthrop	426	407	200	405	379	268	13	...	680	794	606	506	4,684	74
Rose, Ayr.	276	185	71	...	...	70	881	772	651	591	493	424	4,384	94
Jenny, "	569	516	551	592	565	499	421	333	221	33	793	751	5,844	30
Jenny, 2d, Ayr.	215	203	255	240	214	171	14	...	145	484	389	342	2,672	79
Fanny, Ayr.	813	668	733	637	607	522	422	350	272	121	601	556	6,305	25
Tilly, "	...	...	798	1,117	1,021	910	796	687	587	456	206	...	6,578	97
Ops, "	...	...	694	802	883	794	667	595	534	496	429	364	6,348	67
Twinney, "	...	70	835	740	757	679	602	558	525	440	367	339	5,912	56
Drusilla, "	...	...	259	889	936	807	713	634	619	558	452	397	6,264	70
Queen, "	...	148	1,162	1,043	1,031	958	846	809	759	737	604	499	8,596	55
Edna, "	...	...	860	826	857	743	662	567	528	477	392	328	6,220	60
Ozora, "	...	...	550	1,068	1,060	887	770	684	627	597	478	455	7,176	73
Mona "	...	...	...	...	...	309	667	560	432	431	347	337	3,086	166
Selena "	...	...	29	773	815	751	667	721	538	538	477	431	5,740	88
Total......18														
Average													5,768	67

N. B.— Five natives, the rest Ayrshires. Jennie 2d a heifer. From the close of this year all our stock has been Ayrshire.

WAUSHAKUM MILK RECORD.

1871.
WEIGHT OF MILK EACH MONTH, IN POUNDS.

NAMES OF COWS.	Jan.	Feb.	Mar.	April.	May.	June.	July.	Aug.	Sept.	Oct.	Nov.	Dec.	Total in Year.	No. days dry.
Jessie............	....	....	471	592	505	437	375	333	326	297	226	127	3,689	86
Jenny, 2d........	332	325	409	337	290	257	209	181	149	34	116	69	2,708	53
Selena...........	360	298	111	....	....	464	719	704	664	712	525	510	5,067	88
Queen of Ayr....	344	44	672	953	980	747	557	651	644	601	502	440	7,135	26
Edna.............	143	....	....	508	949	797	591	556	475	518	342	270	5,149	83
Tilly.............	428	935	970	785	780	630	483	581	138	....	....	....	5,730	129
Rose.............	330	88	....	....	718	692	578	588	504	436	160	59	4,153	89
Fanny............	484	470	533	459	456	434	423	429	428	400	322	309	5,147	....
Jennie...........	647	570	659	567	523	489	234	....	212	753	577	223	5,454	77
Ops..............	215	261	882	708	689	612	543	558	519	465	347	373	6,172	19
Mona.............	280	100	....	....	....	616	709	631	453	356	266	228	3,639	104
Ozora............	338	8	....	....	381	980	909	892	726	623	480	517	5,854	107
Drusilla.........	348	472	651	566	576	489	446	447	389	375	208	83	5,050	....
Twinney..........	114	....	356	936	967	788	645	644	543	395	41	....	5,429	107
Total........14												Average	5,027	69

N.B.— All Ayrshires.

1872.

NAMES OF COWS.	Jan.	Feb.	Mar.	April.	May.	June.	July.	Aug.	Sept.	Oct.	Nov.	Dec.	Total in Year.	No. days dry.
Twinney	67	849	856	748	802	712	618	544	633	565	481	418	7,293	28
Drusilla	12	...	41	879	930	771	639	588	680	571	403	142	5,656	84
Ozora	522	411	305	7	...	...	879	906	847	718	547	494	5,636	91
Mona	180	41	...	...	...	129	783	790	702	635	465	388	4,213	127
Ops	395	344	271	71	...	...	217	971	979	887	764	672	5,571	97
Fanny	260	149	12	...	...	942	1,039	909	926	834	670	559	6,299	92
Tilly	...	121	935	868	863	695	607	579	618	389	...	...	5,675	120
Edna	151	...	211	984	876	793	707	629	643	626	490	465	6,525	53
Queen of Ayr	362	174	447	511	551	481	439	446	404	369	325	337	4,846	...
Selena	463	368	371	382	191	...	841	817	727	585	439	348	5,532	38
Georgie	542	435	471	460	420	110	955	837	889	837	644	527	7,127	19
Model of Perfection	682	591	611	519	534	459	630	569	643	618	493	461	6,810	...
Lady Kilbirnie	30	...	...	824	1,159	1,036	844	777	900	734	604	521	7,429	95
Total 13													6,047	65

Average.............. N. B.— All Ayrshires

WAUSHAKUM MILK RECORD.

1873.
WEIGHT OF MILK EACH MONTH, IN POUNDS.

NAMES OF COWS.	Jan.	Feb.	Mar.	Apr.	May.	June.	July.	Aug.	Sept.	Oct.	Nov.	Dec.	Total in Year.	No. days dry.
Twinney	346	246	44	161	957	963	800	759	724	604	446	404	6,454	42
Drusilla	...	...	791	609	565	684	594	475	472	464½	330	274½	5,259	59
Mona	310	240	92	...	...	...	...	751	833	689½	501	446½	3,863	137
Ops	600	559	590	422	498	589	336	188	170	1,059½	837½	763½	6,612½	25
Tilly	...	676	808	603	686	777	607	663	654	494½	302	124½	6,395	35
Edna	373	312	355	305	341	332	221	84	131	865	775½	715½	4,810	30
Queen of Ayr	297	234	90	...	304	1,165	870	826	718	687	625½	558	6,474½	69
Selena	303	253	256	222	247	52	...	...	...	...	...	...	1,333	207
Georgie	483	405	415	243	150	13	...	1,116	1,110	787½	657	715	6,094½	68
Model of P.	334	117	44	...	...	756	1,000	1,022	967	771	612	613½	6,192½	106
Lady Kilbirnie	330	166	...	...	...	1,338	1,070	1,036	1,008	844	642	646	7,124	82
Isobel [17]	...	499	753	574	661	680	540	548	570	488	403½	380	6,076½	40
Sea Bird [18]	...	...	...	...	...	...	...	...	487	709½	450½	500	2,147	...
Olee [18]	...	...	...	...	...	...	...	...	395	666	559½	560½	2,181	...
Total													5,436	74

Average 12.6

[17] Isobel is a heifer. Calved Feb. 9.
[18] Sea Bird and Olee are heifers. Calved Sept. 11 and 12 respectively.

1874.

Names of Cows.	Jan.	Feb.	Mar.	Apr.	May.	June.	July.	Aug.	Sept.	Oct.	Nov.	Dec.	Total in Year.	No. days dry.
Drusilla	232½	161	55½	...	456	650½	539	512½	503	481½	320	275	4,186½	38
Tilly	19	...	854	840	880½	725	500	532½	481	282½	25½	...	5,140	104
Edna	680½	544½	569½	501	559½	565	412½	308½	151½	...	508	887½	5,688	47
Georgie	708	606½	617½	444	399½	292½	139	1,417½	1,329½	1,070½	646	620½	8,271	22
Model of P.	521½	386½	275	678	780	791	586	638	613½	477½	432	415	6.594	...
Lady Kilbirnie	664	470½	311	23½	...	...	...	1,097½	839	654	434	402	4,895½	111
Sea Bird	480½	445	473	386	439	472	328	280	87½	520½	510½	527	4,949	14
Sally[19]	...	...	...	...	...	...	...	...	236	519½	423	421	1,599½	...
Total.........7½	Average.........												5,661	49

[19] Sally, a heifer. Calved Sept. 15.

The animals whose yields are represented, it will be remembered, were selected with our best judgment, for the best of natives and for the best of Ayrshires. We have, therefore, in these yields, a basis for ascertaining the comparative value of the natives and Ayrshires of similar grades and under similar management, the variation of the same cows in different years, and other matters of interest, in a reliable form.

The natives were kept for their milk alone, and none of the calves were raised. The Ayrshires are a breeding herd, and not only are the calves raised, but the supply of food is regulated with great care, in order to avoid the hazard arising from high feeding for milk.

It is seen that, reducing all the figures to the basis of one year, we have for the average yield, —

```
68   Native cows,   4,605 pounds, or 2,141 quarts.
67.9 Ayrshire  "    5,550    "       2,581   "    (including heifers.)
 3   Jersey    "    2,506    "       1,119   "
```

Perhaps arranging our results as in the following table will give a correct showing of the differences between the native and the Ayrshire in percentages of the whole number of cows kept, multiplied by the number of years kept. Thus, a cow kept for three years would appear in this table as three cows kept one year.

SUMMARY. PERCENTAGES.

Annual Yield.				Native.	Ayrshire.	Jerseys.	Native.	Ayrshire.
Under 3,000 lbs.				5	3	1	7.35	4.48
Between 3,000 lbs. and 3,500 lbs.				5	2	1	7.35	2.98
"	3,500	"	4,000 "	9	5	1	13.23	7.47
"	4,000	"	4,500 "	10	6		14.70	8.96
"	4,500	"	5,000 "	11	4		16.17	5.96
"	5,000	"	5,500 "	6	8		8.82	11.93
"	5,500	"	6,000 "	11	12		16.17	17.92
"	6,000	"	6,500 "	3	13		4.41	19.40
"	6,500	"	7,000 "	4	6		5.88	8.96
"	7,000	"	7,500 "	2	6		2.94	8.96
	7,500	"	8,000 "	1	0		1.47	—
"	8,000	"	8,500 "	1	1		1.47	—
"	8,500	"	9,000 "	0	1		—	1.49
				68	67	3	99.96	99.51

Or, grouping on a larger scale,—

	Natives.	Per cent.	Ayrshires.	Per cent.
Under 4,000 lbs.	27.93	} 58.80	14.92	} 29.84
Between 4,000 and 5,000 lbs.	30.87		14.92	
Between 5,000 and 6,000 lbs.	24.99	} 41.16	29.85	} 70.15
Over 6,000 lbs.	16.17		40.30	

Or, 68 Native cows' yields, 34 different cows = 325,023 lbs. total yield.
    67 Ayrshire " 18 " " = 368,884 " "

As a constant process of selection was continually in progress with the native herd, let us place side by side the 34 Natives and 34 Ayrshires yields.

34 best Ayrshire yields,	225,063 lbs.	Per cow,	6,620 lbs.
34 best Native "	199,877 "	"	5,878 "
34 poorest Ayrshire "	152,822 "	"	4,494 "
34 poorest Native "	125,146 "	"	3,680 "

The average for the three years when Natives were principally kept was 4,562 lbs.; for the five years of Ayrshires, 5,588 lbs.

		Per day per cow while in milk.	Per day per cow per year.
1867.	All Natives,	16.81 lbs.	12.8 lbs.
1868.	Principally Natives,	16.06 "	13.24 '
1869.	" "	14.31 "	11.45
	Average,	15.72 "	12.5 "
1870.	Ayrshires principally,	19.35 lbs.	15.8 lbs.
1871.	"	16.98 "	13.77 "
1872.	"	20.59 "	16.94 "
1873.	"	19.12 "	15.43
1874.	"	19.96	15.51 '
	Average,	19.20 "	15.49

Thus the "Ayrshire years" show a yield of about 1,200 pounds more per cow in milk than do the "Native years."

The "Ayrshire years" also show a yield per cow for the year of 1,095 pounds more than do the "Native years."

These statistics, with every feature in favor of the native cow, certainly justify claims for high value to the Ayrshire stock as milkers.

BUTTER.

Although the making of cheese has been carried on in Ayrshire from a remote antiquity, it has not excluded the practice of using the milk, at least since the beginning of the present century, for other purposes. The manufacturing industries of this region have concentrated population and fostered artificial wants. Previous to the year 1811, and probably very much earlier, butter was manufactured from the milk in winter, but in a ruder method than at present. As early as 1811, Aiton could state that all the milk made at more than a mile and a half, and not more than ten miles from Glasgow, was converted into butter and sold in that city.

In 1869 we ourselves found butter made extensively in the dairies throughout the county; and in all the cheese dairies that came under our observation, the Sunday's milk was reserved for the making of butter.

In 1864 Mr. R. J. Thomson, of Kilmarnock, tried a series of experiments on feeding roots to Ayrshire milch cows. The percentages of cream varied from $12\frac{1}{2}$ to $14\frac{1}{2}$ in the four animals, as the average of a six weeks' trial. In another trial by the same gentleman in 1865, with 8 cows, the cream percentage varied from 9 to 16, and the average was $13\frac{1}{3}$ per cent. In still another trial with four animals, the result was about 12 per cent. These percentages were read off after standing 24 hours.[1]

In America we have but few records of the cream percentage. Mr. Thomas Miller, of Delaware County, New York, writes us that his cow, Favorite, gave 25 per cent of cream in 1871.

The result of numerous trials on Waushakum Farm gives a variation of from 9 to $18\frac{1}{2}$ per cent. We assume the average to be about $14\frac{1}{2}$ per cent.

Ro. Forsyth,[2] writing in 1807, states that 8 Scottish pints of milk on the average produce a pound of butter of 22 ounces. This is in the proportion of 1 pound of butter to $22\frac{4}{10}$ pounds of milk.

Aiton,[3] in 1814, states that 2,453 wine quarts of milk produced 228 pounds of butter, — a proportion of 1 to $23\frac{3}{10}$ pounds.

A farmer in Stirlingshire,[4] quoted by Mr. Colman, gives his proportion as a pound of butter to 16 quarts of milk.

Professor J. Wilson[5] gives the proportion of a trial as 1 to 20.

In 1830, in Ayrshire,[6] 12 cows gave during one week, 1,075 quarts of milk, which produced 84 pounds of butter. This is in the proportion of 1 to $27\frac{7}{10}$ pounds.

An experiment at the King William's Town Dairy,

---

[1] Prize Essays High. Soc. 1868–9, p. 52. [2] Beauties of Scotland, iii, 77.
[3] Sinclair's Scotland, iii, 65. [4] Farmers' Lib. iii, 306.
[5] Journ. R. A. S. 2d ser. iv, 320. [6] Prize Essays H. Soc. 2d ser. ii, 253.

in 1839, gave a pound of butter to each $9\frac{1}{2}$ imperial quarts of new milk.[7]

In Derbyshire[8] a cow in pasture, giving 20 quarts of milk, produced 34 ounces of butter, or in the proportion of 1 pound to $9\frac{4}{10}$ quarts.

Magne[9] gives a table representing the results of trials by different farmers in Ayrshire, as follows:—

Mr. Burnet, Gadgirth........25 gallons of milk, give 8 lbs. of butter.
Mr. Alexander, Southtree....$22\frac{1}{2}$ "   "   "   9 "   "
Mr. Rankins 69   "   "   " 24 "   "
Mr. Buttery, at Calder Bank, 6 Scotch pints of milk give 1 pound butter.

The proportion as indicated here is 1 pound to 15 quarts, except in the case of Mr. Buttery, where we have 1 pound of butter to every $23\frac{4}{10}$ pounds of milk.

We have also a few experiments made in America. The cow Swinley,[10] imported in 1839, furnished in 4 days 102 pounds of milk, which made 5 pounds of butter. This is the proportion of 1 to $20\frac{4}{10}$ pounds.

Mr. E. P. Prentice[11] is said to have had a cow which gave 118 pounds of milk in three days, which produced 9 pounds 5 ounces of butter,—a proportion of 1 pound to $12\frac{6}{10}$.

Mr. H. S. Collins[12] gives his proportion as 1 pound butter to $8\frac{3}{4}$ to 10 quarts of milk.

Mr. Allis, in the Report of the Agriculture of Massachusetts for 1871-2, makes a statement of 60

---

[7] Journ. R. A. S. 1. 413. [8] Johnston's Ag. Chem. p. 537.
[9] How to Choose a Milch Cow, pp. 136, 139.
[10] Farmers' Lib. iii, 305.   [11] Count. Gent. July 23, 1853.
[12] Report of Conn. Bd. of Ag. 1867, p. 146.

BUTTER. 51

ounces of butter from 20 quarts of milk, — a proportion of 1 to $11\frac{2}{10}$ pounds. In the same volume is a statement of a proportion of 1 to $17\frac{8}{10}$ pounds.

Mr. F. H. Appleton writes that his cow Maud yielded butter in the proportion of 1 to $15\frac{6}{10}$ pounds.

Four experiments carried on at Waushakum Farm, by churning small quantities of milk in a bottle, give a proportion of 1 to 25 to 28. This, however, does not give a true result except as marking a limit, for the trials were not designed to obtain an answer to this question. A portion of excellent Jersey milk, churned about the same time, yielded a proportion of about 1 to 40. From incomplete experiments we should place the proportion for a fair herd of Ayrshires at about 1 to 20.

Mr. Colman,[13] while travelling through Scotland, was told by a farmer in Stirlingshire of the highest eminence that his Ayrshire cows, in the best of the season, averaged one pound of butter per day, and that he had known two Ayrshire cows to make 2 pounds 2 ounces each per day.

Mr. Caird[14] speaks of these cattle being kept in Norfolk County, England, for the purpose of making butter for the London market.

Mr. Robert Wilson, of Kilbarchan, writes us that he had owned a cow that gave 2 pounds 6 ounces daily, or 16 pounds weekly; and another that did the same on two trials in two successive years.

In the experiments on the food of animals, made

---

[13] Farmers' Lib. iii., 306. [14] Caird's English Agriculture in 1850-1, p. 170.

by Dr. Thomson,[15] the two cows experimented on gave 11¼ pounds and 8 pounds in two weeks in June.

Jean Armour,[16] the well-known cow of Mr. Peters, gave 6 pounds 3 ounces of butter in 3 days in July.

The cow Swinley,[17] imported in 1839, gave, in April, 43 pounds 6 ounces of butter; in May, 42 pounds 4 ounces; in June, 44 pounds 7 ounces; total in three months, 130 pounds 1 ounce. After June, her milk was not kept separate from that of the herd. Largest yield for one week, 14 pounds.

A cow owned by E. P. Prentice,[18] of Albany, gave 12 pounds 7 ounces of butter on grass feed.

One of the cows imported by the Massachusetts Society for Promotion of Agriculture,[19] gave in the winter trial 10 pounds of butter a week.

Mr. Miller, of Delaware Co., N. Y., writes us of one of his cows giving 14 pounds 13 ounces of butter in one week in June; also of another giving 14 pounds 11 ounces, and of a third which gave 13½ pounds in one week in July, 1865, and 18½ pounds the same week in 1867.

Mr. Charles Shepard, of Ogdensburg, N. Y., writes us that one of his heifers yielded 14 pounds butter in a week, and that one aged cow gave 18$\frac{3}{16}$ pounds in one week.

---

[15] Thomson's Food of Animals, N. Y., p. 55.
[16] H. H. Peters' Cat. 1865.
[17] Farmers' Lib. iii, 305. Trans. N. Y. Ag. Soc. 1842, p. 264.
[18] Trans. N. Y. Ag. Soc. 1851, p. 413.
[19] Farmers' Lib. iii, 304.

A cow owned by Mr. A. S. Lewis,[20] of Framingham, Mass., gave $12\frac{1}{4}$ pounds of butter in a week in September.

Two trials only, on cows at Waushakum Farm, resulted in 1 pound and $1\frac{1}{16}$ pound a day. This was in October.

---

[20] Ag of Mass. 1853, p. 299.

CHEESE.

Although the Ayrshire is universally acknowledged to be a large producer of cheese, yet we find very few exact observations on record.

Ro. Forsyth,[1] about 1805, states the proportion as 70 Scotch pints of skimmed milk producing a stone of marketable cheese, and 53 pints of new milk during the season. This is a proportion of 1 to $8\frac{6}{10}$ pounds, if the Ayrshire stone is the weight referred to.

Aiton,[2] in 1811, gives the proportion as from 50 to 55 pints to the stone of 24 pounds of sweet milk cheese. This is a proportion of about 1 to 8 or 9 pounds.

Again, in 1814,[3] he states that the usual estimate is that 55 pints of milk give an Ayrshire stone of cheese. This is a proportion of 1 to $8\frac{9}{10}$ pounds.

In a reference to Dunlop cheese[4] the proportion is again stated as 1 to $9\frac{6}{10}$ pounds.

Magne[5] gives a table of the estimates of various farmers in Ayrshire, as follows:—

Mr. Alexander, Southbar.........$22\frac{1}{2}$	gals. milk give	24	lbs. cheese.
Mr. Sanderson, Blackcastle ......$26\frac{1}{4}$	" " "	$27\frac{1}{2}$	" "
Mr. Wm. Peats....................23	" " "	24	" "
Mr. James Peats ....$23\frac{85}{100}$	" " "	24	" "
Mr. Ranburn....................$24\frac{58}{100}$	" " "	24	" "

---

[1] Beauties of Scotland, iii. 77.
[2] Survey of Ayrshire, p. 466.
[3] Sinclair's Scotland, iii. 69.
[4] Journ. of Ag. 1st ser. vol. v, p. 363.
[5] How to Choose a Milch Cow, p. 139.

This would establish the proportion as 1 pound of cheese to each 3.872 quarts of milk, or as 1 to $8\frac{1}{4}$ or 10.1 pounds, according as wine or beer measure is intended.

About Kilmarnock, in 1869, we were informed by intelligent farmers that 3 pounds of curd a day per cow was considered a large yield, but $2\frac{1}{2}$ pounds per day was about the usual quantity, taking the average of all the cows.

Archibald Sturrock,[6] in 1866, estimates the yield annually per cow throughout Ayrshire at 432 pounds, and for the season of six months, 384 pounds in the best grazing district, and 288 pounds in the poorest.

A writer in 1872[7] says that cheese is made in Ayrshire from the time the cows go to grass until the commencement of November, and the quantity each cow is estimated to produce is from 3 to 4 hundred weight, or from 336 to 448 pounds.

A letter from Mr. Robert Wilson places the daily production of cheese as 3 pounds for a good average. From 3 to 5 hundred weight (336 to 560 pounds) may be reckoned a cow's produce of cheese, — the higher quantity when the pasture is superior being as possible as the smaller when it is inferior.

The cheese made in Ayrshire is a sweet-milk cheese, mild-flavored and rich, called the "Dunlop." It was begun to be made by some farmers in the Bailliary of Cunningham prior to the middle of the last century, and it has gradually extended over the

---

[6] Prize Essays H. Soc. 4th ser. I, 80.
[7] Milk Journal, Jan. 1, 1872, p. 20.

counties of Ayr, Renfrew, and Lanark, and in Galloway.[8] At the present time there is a considerable manufacture of the English cheddar in these regions.

Cheeses made in Scotland are neither washed nor rubbed nor greased, on the outside, nor painted like some of the Dutch and English cheeses, but merely laid up to dry on clean boards, in a place neither dry nor damp, and frequently turned.[9]

The Dunlop cheese is generally not so acrid in the taste as most of the English cheese, nor is it so hard and dry as that of Holland; it is softer and fatter than either.[9] Subjoined is an analysis, by Mr. Jones, of a cheese made in 1845, and analyzed in 1846,[10] expressed in percentages.

Water	38.46
Caseine	25.87
Fat	31.86
Ash	3.81

"Somewhere about five o'clock, A. M., the morning milking of the cows takes place. The milk is carried direct in the 'luggies' as drawn from the cows, and emptied through a very fine wire-cloth sieve ('the milsey'), or else through a thin canvas cloth, into a large 'milk-boyen' or tub standing in the contiguous diary-room. . . .

"The cream of the previous evening's milk is skimmed off, and the remainder being warmed in a vessel in the boiler to about or fully 100°, is then added through the sieve, along with the cold cream,

---

[8] Journal of Agriculture, 1834-5, p. 358. [9] Same, p. 362.
[10] Journal R. A. S., 1858, p. 420.

to the morning's meal already in the tub, and raising the whole when added to an uniform temperature of from 86° to 88°. Milk, as it comes from the cow, is about 96°. After stirring in the 'rennet,' the milk takes about thirty minutes — seldom less, sometimes more — to properly 'thicken' or coagulate. . . . The breaking of the thickened fluid comes next in course. This is effected, generally, by passing the arm and outspread palm softly and steadily in all directions through the coagulated milk after a short time allowed for the curd to subside, — most assisting by pressing it gently down with their palms, — the whey is lifted off with a suitable vessel, and poured through a sieve into some receptacle for the use of the pigs. The massed curd left in the 'boyen' is then cut into about four-inch cubes, which are tied into a wet, coarse cloth, spread within a square wooden box, with perforated bottom and sides (termed a 'dreeper' or 'drainer'), and subjected to a pressure of about twenty pounds or so. The curd undergoes this process four to six times, with lengthening intervals between, and each succeeding time being cut into still smaller pieces, with increased pressure, till the whey has been as completely expressed as the 'dreeper' is capable of. . . . The broad lump of solid curd . . . is first cut into four-inch cubes or so, and which are then put through the curd-mill, which fractures or tears, rather than cuts the bits into fragments.

"Due allowance of salt having been mixed, in the proportion of 1 to 48, a fit-sized 'chessat' (abbreviation for cheese vat) is selected, and a cheese-cloth being spread within it, the prepared curd is firmly pressed in with the hand, the corners of the cloth being brought up over all, and the contained curd, it may be, jutting some three to four inches above the edge of this chessat. By this time it is rather past noon of the day. Some then place the chessat in front of the kitchen fire, with the lid weighted, and standing there for most of the afternoon, frequently turned so as to equalize the heat, and at evening it is put in the cheese-press. Others warm the prepared curd in a vessel before the fire prior to making up the cheese. During the process of pressing, too, the chessat is occasionally brought to the kitchen fire. . . . A certain degree of heat, tending to improve the quality as well as facilitate the pressing, must be kept up within the curd whilst becoming solid. . . .

"The made-up cheese we put to press towards evening is taken out of the chessat on morning of second day, and is then — in very many dairies, though not by all — scalded with the cloth on for near an hour in hot water fully as hot as can be tholed with the hand. It is wiped when taken from the hot bath, wrapped in a dry cloth, and put to press again. It is removed and dry cloths substituted at noon and evening of same day, reversing the cheese in chessat at each remove. Like performance has to be gone through, — it may be only once in some dairies, perhaps twice in others, and even three times occasionally, on the third day, by which time the cheese is perfected. The dairy-woman has thus always three cheeses in hand. The cheese is then placed, without more ado, wherever it is to lie, till sold and sent off, being reversed and rubbed with a dry cloth every day for a short time at first, and afterwards at lengthening intervals. None of their inward coloring with annatto, or outside painting with Spanish brown; nor sweating nor greasing nor canvas swaddling at all. Just the naked, unadulterated truth."[11]

---

[11] Archibald Sturrock, in Prize Essays H. Soc. 1866-7, p. 89.

MEAT.

In Scotland, the older cows and the steers are used extensively for the purpose of food. Almost every reference to the merits of the Ayrshire breed refer to their grazing qualities. Thus Aiton says that their beef is better than that of most other breeds on account of the fat being more evenly mixed with the lean, and claims that the dry cow fattens *faster* than any other breed.[1] Colman quotes an Ayrshire farmer who claims there are no better feeders, and that when fatted their beef is as good as that of the West Highland breed.[2]

A reference in the "Dumfries and Galloway Courier" says that there are many instances in which Ayrshires of the same age and size with Galloways have attained to a nearness kindred weights. Two-year-olds of this breed will give the same price as Galloways of the same age.[3]

Sinclair says that they fatten faster and to as great an extent as any of the other breeds in Scotland;[4] and G. Murray (in Jour. R. A. S. of England) says that they are of a kindly disposition, and feed readily when tied up in the stall or put in good pasture.[5]

---

[1] Survey of Ayrshire, 429.
[2] European Ag. ii, 318.
[3] July 11, 1842, quoted in Journ. of Ag. xiii, 1st ser. p. 228.
[4] Code of Agriculture, p. 19, note 142 of notes.
[5] Journ. R. A. S. 1866, p. 56.

H. N. Fraser, on the contrary, denies their value for feeding purposes, and says that they are of slow maturity.[6]

Quotations giving opinions of their value for grazing could be indefinitely multiplied. The truth seems to be, that the shapes are not those which are most profitable to the butcher. They cut up with not such economy as the Short-horn, nor do they arrive as early at maturity. As feeders they are the equals of many breeds used for grazing, when rightly treated, but have not the same aptitudes which have been bred so especially in the Short-horn.

In quality of meat, they can hardly be excelled; our experience in Glasgow and Ayr in 1869, and with a barren heifer in 1872, justifies us in describing their meat as fine-grained, high-flavored, juicy, and marbled with fat.

A few extracts from the catalogue of H. H. Peters, of Southboro', Mass., will illustrate the capabilities of this breed, among the hills of New England, as possible beef producers. "The imported cow Ada, proving barren, was fattened during the winter of 1862-3. About the first of April, 1863, she was slaughtered. Her dressed weight was 1,009 pounds, of which the beef weighed 882 pounds, the tallow 111 pounds, and the hide 70 pounds. The quality of her beef was pronounced, by persons well qualified to judge, superior. It was fine-grained, and the fat and lean so well mixed as to produce the marbled appear-

---

[6] Pr. Essays H. Soc. 1868-9, p. 331.

ance which is highly prized by epicures. The meat was also in large quantity in proportion to the bone.

"The imported cow Nannie, nine years of age, dropped a calf in September, 1862; was milked until July, 1864, when she ran in a short pasture until November, without extra feeding; since that time she has had meal, eating most heartily, and increasing in weight more rapidly than grade Short-horns which have been fed with her. She weighed 1,372 pounds in March, 1865.

"A full blood steer, three years old March 5, 1865, now weighs 1,332 pounds, and girths six feet and ten inches; had never tasted meal until the middle of November last, four months ago."

OPINIONS OF THEIR WORTH.

As early as 1805 the merit of the breed seems to have been known beyond their home, and Ro. Forsyth[1] mentions their presence and estimation in Renfrewshire, Perthshire, Dumbartonshire, and Stirlingshire, and Aiton[2] mentions their inroad into Galloway in 1802. In 1842 the "Dumfries and Galloway Courier"[3] speaks of them as "creeping fast over Dumfriesshire and Galloway." In 1872 they had been around distant Inverness for a number of years.[4] Their merits have also been recognized in foreign countries, as witness their exportation to America, the Canadas, France, Oldenburg, and Norway.

Sinclair[5] writes, "The Ayrshires are perhaps the best milkers of their size in Great Britain, and at the same time are excellent feeders when dry of milk, for they fatten faster and to as great an extent as any of the other breeds in Scotland."

The "Dumfries and Galloway Courier"[3] of July 11, 1842, says, "The opinion is becoming more and more general that the Ayrshire breed of cows is superior to any other in our island, qua the pasture, the byre, and the milk-house. In size and weight they suit the

---
[1] Beauties of Scotland, iii, 8, 347, 405; iv, 245.
[2] Survey of Ayrshire, p. 426.
[3] July 11, 1842, quoted in Journ. of Ag. xiii, 1st ser. p. 228.
[4] Pr. Essays H. Soc. 1872, iv, 51.
[5] Code of Agriculture, p. 19, note 142 of notes.

grass enclosures of Scotland, but especially of such districts as Ayrshire, Lanarkshire, Dumfriesshire, and Galloway, where such herbage as best suits dairy stock abounds. They are easily fed, and in proportion to bulk give more milk than any other. Already, as milkers, they have supplanted to a great extent all the other kinds in the county from which they take their name. . . . Galloways, as beefers, are excellent stock, but we have known many instances in which Ayrshires of the same age and size obtained to a nearness kindred weights. Two-year-olds of this breed will give the same price as Galloways of the same age."

In 1837 Baron Malzahn Sommerstorff, on the part of an association in Pomerania, imported 185 cows, and he testifies that he had found no breed that gave so much milk upon moderate food as the Ayrshires.[6]

At the Universal Exposition at Paris, "pre-eminently did the Ayrshires and Alderneys stand first in the first division, and the former received the impress of the approval of the foreign agriculturist by the rapidity with which they were bought up, — a rapidity unequalled by that of any other breed, excepting the Bretons. . . . In reference to the division of the different breeds of cattle we have given above as milkers, we may state it agrees with the results which have been obtained at the Imperial School of Grignon, from carefully conducted experiments. The Ayrshires are proved there to give the largest quantity of milk in proportion to the quantity of food

---

[6] Alb. Cult. Jan. 1, 1844.

consumed, the Swiss cattle the next, and the Bretons next."[7]

"M. Bonnemant, fully appreciating the valuable milking qualities of the Ayrshires, and their suitability for Brittany, has introduced a considerable number of first-rate animals of that breed."[8]

Mr. Horn, before an English Farmers' Club, proceeds, "I next advert to the Ayrshires, and I believe, taken as a breed, they are the most select as to milking properties. . . . I hesitate not to state that we have no other class of cows, taken as a breed, that will produce the quantity of milk for food consumed. Hence the high estimation in which they are held in cheese-making districts."[9]

Dr. Voelcker,[10] the honored chemist of the Royal Agricultural Society of England, says, "For dairy purposes in cheese districts, the Ayrshires are justly celebrated; indeed, they seem to possess the power of converting the elements of food more completely into cheese and butter than any other breed. The food in their system appears to be made principally into milk and not into meat."

G. Murray,[11] an English writer, also states that "this breed stands unsurpassed for the purpose of the dairy, and has within the last twenty years been much improved with special reference to its milking capabilities; they are of a kindly disposition, and feed readily when tied up in the stall or put on good pasture."

---

[7] Journ. of Ag. 1855-7, vii, 417. [9] Gard. Chem. and Ag. Gazette, Sept. 19, 1863.
[8] Journ. of Ag. 1857-9, viii, 253. [10] Journ. R. A. S. of Eng. xxiv, 308.
[11] Journ. R. A. S. of Eng. vol. 2, 1866, p. 56.

John P. Reynolds,[12] the Commissioner from Illinois to the Universal Exposition at Paris, in 1867, in reporting upon the horned cattle there exhibited, writes, "At the Imperial Model Farm of Vincennes, where one hundred cows are kept for milking, and the sale of their product in Paris, the varieties are Ayrshire, Brittany, Swiss, Normandy, and Flemish, which, as M. Tisserand informed me, taking into account the food consumed, rank for quantity of milk in the order I have named them."

H. N. Fraser,[13] in a prize essay, writes, "Dairies being very numerous in Dumfriesshire, Ayrshire cattle occupy the most prominent place, cows of this useful and valuable breed being considered the best milkers, and at the same time easier kept than any other."

A correspondent of the "Country Gentleman" from Passaic Co., N. J., writes, under date of July 8, 1869, "During the last winter I kept over three Ayrshire cows and three common ones, fed them all alike, and in the spring the Ayrshires looked fat, smooth, and nice, while the common cattle were poor and ragged, — so bad that I was ashamed of them, while I was proud of the others. Another thing is, they give a great deal more milk, and the milk is as rich as any milk."

Mr. H. S. Collins, of Connecticut, speaks of this breed "being kept on his farm with grades and natives, fed and treated precisely alike winter and summer; the Ayrshires have proved the most hardy, the

---

[12] Trans. Ill. Ag. Soc. vol. 7, p. 696.
[13] Prize Essays High. and Ag. Soc. 1868–9, p. 331.

best milkers, both in first yield and in holding out, have kept in the best condition on the same food, and have finally superseded the others by their own merits."

Mr. Charles Shepard, writing us from Ogdensburg, N. Y., says, "It is admitted by all dairymen in this section, that wherever the Ayrshire blood prevails in the herd, that cow winters best and produces most on short feed."

Mr. Edward L. Coy, Washington Co., N. Y., writes us: "In fact, I never had my natives keep in as good condition, both summer and winter, on the same care and feed as my Ayrshires do."

Mr. J. D. W. French, of North Andover, Mass., writes, "As compared with grade or native stock, I find the Ayrshires hardier and easier to keep under the same treatment"; while Mr. F. H. Appleton, of West Peabody, states, "By experience I call my Ayrshires very hardy naturally. But I think that they can be made tender or hardy according to the treatment they receive. This is the result of observation among numerous other herds."

Mr. A. P. Ball, writing from the Province of Quebec, testifies, "They have stood alongside of as good grade cows as I had, also by *thoroughbred* Shorthorns; they are easier kept and come out better in the spring on the same description of food than either of the first named. I do not say on the same quantity: of course, they, being smaller than the others, would not naturally require as large a quantity, but I state the same quality, giving each breed what they require and could cleanly consume without waste."

From H. W. Eddy, Watertown, N. Y., we have the following: "My Ayrshires are intelligent, ambitious, and industrious feeders universally; will recognize a stranger instantly; have never had a sick one except when hurt or injured in somewise; and the peculiar fineness of their nature and ambition will compel them to be upon their feet as long as strength holds out. They feed rapidly and earnestly when in pasture; much more of their time is spent in hunting around fence-corners, stumps, and other obstructions for green and sweet food, such as is commonly overlooked by the native cow. In consequence of their intelligence they will resent an injury and appreciate a kindness, making it very necessary to treat them justly."

And with the admirable observations of Mr. Eddy we close our chapter.

ADAPTABILITY.

When the dairyman is invited to examine a breed of cattle new to him, and is asked to substitute such in the place of those with which he has been long familiar, his inquiries will be directed to two aspects of the proposition: first, as to the excellence of the new breed; and second, as to its adaptability to meet the requirements of his situation.

The fact should not escape us that all breeds will not show to equal advantage when brought into the same locality. In nature we observe a nice fitness of the animal for its place. In domestication we also observe this predominance of nature, — the yielding of the animal to more closely fill her environment. Water not more certainly seeks its level than do our animals, domesticated and wild, seek to correspond with the conditions within which they are placed.

Bring the life of the tropics into our wintry clime, and how soon is death around! Bring that of the more temperate zone, there is less of death; but with the survival of life in the species, there is not a continuance of the shades of character and resemblance to its own. For a while divergence obtains, until in length of time equilibrium is restored, and the breed, remoulded, is uniform as before.

When a breed of cattle is introduced to a locality to which it is a stranger, the first exertion of its force is spent in seeking an adaptation to its new environment. The result may be a retention of its own traits, or it may be the loss in part or wholly of certain characters and the acquirement of new ones.

Suppose the improved Short-horn to be placed upon the plains of Texas and to be left quite to themselves. If they survive the change, and you seek them after the lapse of time, will it be the Short-horn, or even the old Yorkshire, that you will find? Will you not find more of horn and less of body, a greater length of leg and a build for travel? Will there not be more speed and less fat? Will not the type conform to the conditions within which they are placed?

The dairyman should consider these things. Yet the artificial conditions to which his cattle are subjected make the intention of nature less apparent. The resisting force is greatest when art is the most upheld, and nature, its power usurped, only slowly and perhaps almost imperceptibly intrudes to thwart her purposes.

The Ayrshire cow, removed to England, is said not to maintain her dairy qualities at the best; there is tendency to flesh. The American-bred Jersey shows more horn, larger bone, and a less deer-like form than the Jersey-born. Have we not seen the Short-horn brought from England, cultivated here for a few generations, and returned, an improvement over the English-bred?

Changes induced by change of environment, illustrations of which are numerous, ought to teach us not to expect the Ayrshire to be always, or often, when taken to new countries, true to her fame. We believe her to be a smaller milker in New England in general than she is in Ayrshire. The atmosphere here carries habitually less water, and there is less of nutritious food in our pastures, and more wear of life in obtaining it. Yet she is a larger milker in New England, we believe, than is any other breed. Although our climate is unlike that of Ayrshire, and our feed less milky, the sum of her conditions offers not as great contrasts as obtains when the Holstein is sought to be acclimated with us.

The degree of hardiness of a breed may be inferred from the nature of its home. The Ayrshire is exceptionally hardy. Though you may not expect to freeze her blood in the yard, and at the succeeding thaw find her milk flow unimpaired, her coat sleek, and her back straight, yet she will be as profitable with those who expect all this from a cow as any other. But for her sake we should advise such not to breed Ayrshires. A pump with valves at the bottom of the well will be better property.

Expose her to hard fare and rough winds, she will not be handsome, but the constitution she carries with her, and her inherited vigor, will be manifest. With the bestowal of better feed and reasonable protection, the dormant forces of her nature are awakened, and in uses she acts, and in appearance she looks, the beauty of the yard.

The hardiness of the Ayrshire, her instincts, lending both boldness and prudence to her character; her liveliness of movement; her medium size, and her character for seeking her food, all adapt her to be useful, where many breeds would be out of place, and would be spending their force in contention with adverse circumstances. The Ayrshires show their superiority the most where disadvantages are to be overcome. Scant pasturage, steep hill-sides, sudden changes of temperature, and transitions from a moist to a dry atmosphere, are not favorable to dairy interests. But among such conditions, as in New England, the cow may be as much a necessity as in more favored sections of the country. While the Ayrshire may be as well, or better, adapted to afford profit in the favored localities than other breeds, we bespeak specially her superior claims for such half-fertile localities, where as great work is required of the animal in the obtaining of food as in the utilization of it.

We would not be understood to assert that she is unappreciative of the clover-field, where the feed is to her eyes; in such her udder swells to large proportions. The Ayrshires of the western portion of New York State show how kindly she accepts generous fare. The cheese factory to which many of this breed are tributary tells the story of their worth.

Upon soils of great agricultural capacity the dairyman has more breeds from which to select than he who cultivates a soil of less fertility. To the former, the choice is equally open between the larger and the smaller breeds; with the latter, the smaller breeds only can be considered. If a larger animal be adopted than his lands will carry, nature is ever at work to reduce the size, and only pampering care can maintain it, for the land must eventually determine the size of the animal. To work against material forces, rather than with them, is generally most unsatisfactory.

THE IDEAL AYRSHIRE.

The ideal Ayrshire cow is an animal best designed to fulfil the uses for which she is intended, and which at the same time fills the eye as a thing of beauty, completely in harmony between her shapes and functions. Generations of honest endeavor towards an ideal more or less perfect have developed her into her present proportions. In her is united in a completed whole all of those good points which are recognized as indicating milk-giving quality, by all dairymen the world over. The large digestive capacity, the economy of form and capacity of udder, are her most striking features, indicating usefulness, while these are united with a straightness of back, with openness of vertebral joint and comeliness of proportion, a brightness of eye, and that intelligence of expression so attractive to the observer. She has instincts; she knows well her wants; and her frame and her body, her appearance and her functions, are the happy equilibrium between the powers of nature and the powers of art.

To follow out the line of development of the Ayrshire cow, we must commence with the udder, for it is here that the effort after productive power in the milch cow quickly produced a tangible result. As an obvious feature, this organ early showed its relation to uses, and the recognition of this led the breeder to seek at first an increase of its size, and at a later period an economy of its form; and these two united produce a large part of that condition which we call quality. The changes produced in the milk-vessel necessarily occasioned correlative changes in the cow; and our plan is to follow in this line of divergence until we present the completed animal, our ideal of the perfect dairy cow.

The desired udder (and in the best specimens of a cow the udder desired is very nearly realized) is composed of four glands, of which the udder is the sling. These glands are enclosed and separated by a fibrous tissue, which, reflected from the walls of the abdomen, forms a septa and support. These glands are flattened rather than pointed or oval, as in some other breeds, and these, as well as the septa, are noteworthy for the elasticity and tone of their tissues, as well as for their freedom from fat or muscular matter. The udder accordingly should be close to the body, level and broad, and should derive its capacity from the extent of its attachments. The glands being flattened, free from fat, and possessing a tone or milking habit, make little show when not in use, but snugly attached to the abdomen, are covered and concealed by the soft skin of the milk-vessel, so wrinkled and creased and folded as to convey to the unthoughtful observer an incorrect idea of its capabilities for extension. Yet the looseness of the skin, when considered in connection with the distant attachment, the glandular feel and other well-known signs, afford to the trained observer indications of large usefulness.

When filled, the udder should retain its flatness of form, accompanied with a certain squareness of outline. When viewed from behind it should appear broad and deep, extending far back, its attachments loose even to the vulva, and presenting to the sight no hollowness above the glands, no clefts, nor any vacancy between itself and the twists of the thighs. As viewed from the side it should extend well forward, and its skin should merge into the swollen and tortuous milk-veins. No indentation should be seen between the teats, and the hand passed beneath should clearly render sensible the great breadth and flatness, while the eye takes in a levelness of sole corresponding to a line drawn from a point near the brisket, to the hindermost part of the vessel.

The signification of this udder is its harmony with the uses for which it is designed. The breadth of its attachments not only allows the vessel to have large cubic contents with little depth, thus allowing the glands to be in closer proximity to the channels of supply and removal, but necessitates other modifications of structure.

The economy of the position is such as protects the bag in the largest degree from chance injuries, and the animal is freed from the annoyance of the sag in walking on the road, or grazing. As the result of greater nearness to the heart and the lungs, the blood has less distance to traverse in its rounds, and thus the freedom of its circulation is increased. In the human breast the difference in size in favor of the left can only be accounted for by its greater nearness to the heart, the great agent for supply. Furthermore, the temperature of the close udder is retained and maintained by a less expenditure of force, that is, food, than the pendent one, which exposes a larger surface to the air, and places the obstacle of gravity to the flow of the blood.

The teats should be equidistant and at a sufficient distance apart. They are short, apparently from a correlation in structure with the flattened gland. They should be at some distance apart, as indicating the extent of gland, and set evenly, as indicating the evenness of size of the glands; cylindrical, rather than cone-shaped, as this appears to follow from the tone of the tissues and type of the gland of this breed.

The length of the udder is accompanied by length of quarter, and the breadth of udder by breadth of hip, for the bony framework determines the distance of possible attachments. As there seems a correlation in this breed between the breadth of the hip and the distance from the hip to the buttock, with the broad udder, we should expect a long udder. Hence the broad hip and long quarter so universally admired in this cow.

The squareness of the udder in its attachments not only is indicative of capacity, but also of the broad belly which so almost universally belongs to the cow which best digests her food. The workshop of the belly requires abundant room for the storage and transforming of supplies, and this space, which is furnished by the broad hips, and required and indicated by the broad udder, requires strength of loin and back. Hence the Ayrshire cow should excel in this point. The short ribs should be arched but little, and their length should be great, so supporting the skin, as to leave a deep hollow at the flank when the animal is hungry or thirsty, to be obliterated by repletion. This is indicative of the tendency to milk-giving as contrasted with the tendency of laying on flesh; and such are our requirements.

The looseness of attachment to the udder behind is always accompanied by openness of vertebral joint, and this indicates a certain laxity of tissue and vascularity of system.

The breadth of the udder with its proper accommodation not only requires that there should be breadth between the thighs, but that these should be thin and flat at the point which may be technically described as the twist, for it is preferable to have the pressure on the udder from within rather than from without. The round ham and fleshy thigh is more characteristic of the grazing than of the dairy animal.

It appears to be a physiological law that when nourishment flows to one organ or part in excess, it rarely flows in excess elsewhere. We should therefore expect that the great development of the udder in its functional and structural relations would have an influence in checking excessive development of other parts. We accordingly find that the Ayrshire cow when in milk seldom lays on flesh, nay, more a milk-giver by inheritance, she has failed to develop her forward parts in correspondence with the development of her rearmost half. Through the economy of forces, the food is sparingly used for the building up of parts beyond the necessity of the animal, but is directed to giving largeness to the parts that are tributary to her services, and to a direct reappearance in her products. The appearance of lightness forward is caused rather by the absence of unnecessary flesh and the comparison with the hips and flanks, than by any deficiency in the constitutional functions. The demands of the udder are a large supply of blood, which can best be supplied by a healthy heart, not cramped by position, and purified in lungs of ample power. Hence we seek the appearance of vigor as indicated by the absence of hollowness behind the shoulder, by depth through the region of the heart, and by the show of abundant constitution as seen from the front.

As the Ayrshire cow is possessed of these qualities which are of value, we must look for their perpetuation to the needs of the reproductive system, as indicated by the hook bones being wide apart, and by the ampleness of the bony covering, which, to correspond to the parts as already given, would be described as a pelvis long, broad, and straight.

Heaviness of the neck is a masculine characteristic, and it therefore follows that a departure from this type would be desirable in the milch cow. An extremely thin neck has a cowey look, but may, on the other hand, indicate too great delicacy of constitution. The short, thick neck would indicate hardiness, and if overladen with muscle, would suggest barrenness.

To harmonize with our ideal we would choose the neck of medium length, clean and round in the throat, neither too thin nor too thick, and with a symmetrical taper.

The head should have a look of extreme femininity, — a soft, intelligent, motherly expression. To attain this, the nose must be fine and tapering to the ampler muzzle, and the lower jaw neither heavy nor long, especially on the broad-faced type of animal. Breadth of face seems accompanied by a certain refinement of jaw. The lower jaw is stated to be homologous with the fore limbs, and under the law governing the development of homologues, we should expect the fore limbs to be varied in an allied manner. As a matter of fact, the delicate and shapely head is rarely set upon ill-fashioned limbs.

Fineness and flatness of bone and firmness of joint are points which experience has shown to be the accompaniments of thrift in all breeds, and none the less so in a dairy animal. This partiality for economy influences us also in our desire for the diminution of the bulk of those portions of the body that we consider useless. Hence all folds and wrinkles, the large brisket, excessive growth of horn, and all extremes in non-essential points, are discountenanced by the breeder.

The animal may possess all desirable points in detail, and yet be deficient, in that these characters are not so blended as to form an harmonious whole. The totality of structure must be sought, and be so gained, that there be neither unnecessary weight nor bulk, and such an unison of adjustment, as to make the cow an exemplar of economy in its highest type. In so far as the Ayrshire cow has not attained to this, must the breeder strive for it, crowning his art with this success, the tribute to his genius.

The skin is not only the covering to the animal, binding together her parts, and protecting the underlying parts from exposure, but it gives support to the hair, and is studded with glands, and may be considered one of the organs of the animal body, as through it is eliminated not moisture alone, but carbon and other products. Reason as well as experience teaches that its texture and "feel" may be indicative of certain qualities in the animal. It is here that we are enabled to detect degrees of vascularity and thrift; and the hand, taught by experience, can tell by the touch the good from the bad feeder. The skin to be desired in the Ayrshire is neither too thick nor of a papery thinness, but medium. It should be vascular, that is, soft; and although it must not be so well underlaid by fat as in the grazing breeds, it must be loose and easily lifted.

As some of the glands of excretion are homologous with the glands of the udder, their appearance may give us an idea of some qualities of the milk. By experiment we have satisfied ourselves of the relation of the color of the skin secretion as found in the ear and elsewhere, with the color of the butter the animal affords; there is a seeming relation between the glands of the skin and the lacteal gland, which suggests a vicarious action, in a measure, between the two.

The hair is desired to be soft and woolly, on account of its protecting power, thus acting as an assistant in economizing the animal heat. It is probable, however, that the texture of the hair and its form is affected by the climate to a large extent. We doubt whether the woolly hair found on many animals in humid Scotland can be either retained by an imported animal or transmitted to progeny in the dry climate of America. The softness of the hair is affected by food. The oily, unctuous feel of the hair and skin of the animal fed on linseed meal is very perceptible. The functions of the hair are seemingly to protect, and the better it fulfils this purpose, the greater the economy of the animal forces.

The breeder should desire to form the animal in an attractive mould. To do this is to extend their introduction among those to whom the æsthetic is of value. Therefore, fineness of form, as far as is compatible with usefulness, is to be praised. The small horn beautifully curving, the thin ear, the fine tail of good length and well switched, and the color, each and all add value.

Although red and white, or brown and white, are colors towards which many are partial, yet any color but roan is allowable to the thoroughbred. A strong mixture of white, as lending style and adding to their picturesqueness, in our opinion is admirable.

On account of the importance we attach to the presence of the escutcheon on a dairy animal, we have preferred to give the subject a paragraph by itself in this place.

The following language was used by the Committee on Agriculture in their report to the French National Assembly upon "Guenon's Theory of the Milk-Giving Properties of Cows":—

"Admitted by our most learned veterinarians of the Royal College of Alfort and elsewhere, encouraged by the Government, confirmed by a thousand proofs, and sanctioned by your approval, the discovery of M. Guenon may now be considered as having reached the dignity of a science. It applies alike to males and females, to calves and full-grown animals; and from this last fact we may make this fruitful deduction: Hereafter the farmer need rear none but such calves as will make good milkers, handing over to the butcher such as will not."

If Francis Guenon could inspire such enthusiasm and conviction in those appointed to examine into the merits of his claims, what may not be allowed to himself? He says, "It did not suffice to have discovered signs that were characteristic of different sorts of cows; it was necessary to make sure that the same mark might always be relied upon as a positive and certain sign of the same perfection or defect. This could not be effected except by studying a vast number of individuals, by comparing them together, taking into consideration the countries from which they came, their stature, their yield. This was not all; they had to be classed. Conceive what toil this task involved for me, a plain child of nature, who had no idea of such a classification, and found myself under the necessity of establishing one. The endeavor was one to absorb me entirely; I gave up my calling; I travelled about visiting cattle-markets, fairs, cow-stables; I questioned and cross-questioned all who might be expected to know most on the subject, — husbandmen, dealers in cattle, men of the veterinary profession; I became convinced that my discovery had not been anticipated by any one. The marks for distinguishing a good cow from a bad one varied according to the notions of each individual. Some looked to the shape of the horns, others upon that of the udder; some judged by the shape of the animal or the color of her hair; others were determined in their choice by something else: but in these various modes of judging all was vague and uncertain. I became confirmed in the belief that I had made the important discovery of signs that were positive and certain; and in order the better to satisfy myself of the solidity of the ground upon which my method was to rest, I took the precaution to return to the same localities at different times and seasons, that I might trace and ascertain the effects which might attend these variations of nature. All my observations were accurately noted down, and I could at length flatter myself with having acquired a mass of facts which gave solidity and consistency to my system, and imparted the character of positive certainty to that which at first had been but a probable conjecture."

But what are the claims? The Agricultural Society of Bordeaux reported that M. Guenon "has established a natural method by which it is easy to recognize and class the different kinds of milch cows, according to —

"1st. The quantity of milk which they can yield daily.

"2d. The period during which they will continue to give milk.

"3d. The quality of their milk.

"By means of these signs, which are all external and apparent, he has established eight classes or families, which embrace all the varieties of the cow that are to be met with in the different parts of this kingdom. Each of these classes or families is subdivided into eight orders. It is divided, also, into three sections, so that each of the sections comprehends the eight orders."

What are the signs? In Guenon's words, they are the marks "visible upon the posterior part of every cow, in the space embraced between the udder and vulva. They consist of a kind of escutcheons of various shapes and sizes, formed by the hair growing in different directions, and bounded by lines where these different growths of hair meet. The varieties of these escutcheons mark the different classes and orders of cows."

The complete, enthusiastic acceptance accorded the method of Guenon in France, while the author was the presiding genius of his own idea, shows that it has much value. Certain, however, it is, that in England and in our own country the method has not been accepted in all its original elaborateness; and there is a growing conviction that the author saw in the *upward-growing hair* more than others have found and more than is justified by trained experience. That the quality of the cow as a milk-giver is indicated in the escutcheon to a very considerable extent we think must be allowed, but we think it will not answer to read there — as Guenon claim to read — the number of pints of milk each cow would give daily, and much less in precise terms, the quality of the milk.

The internal functions of the cow do not find their complete expression, their tell-tale, if we regard the animal in only one of its aspects. In reason, we ought not to find her record concentrated within the compass of a few square inches, when appetite, constitution, size, and breed must each and all exert an influence in determining yield and quality.

Yet in the mark of the escutcheon we recognize much value. We do not remember to have seen a very good cow that had a small escutcheon, while never have we seen a cow with a good escutcheon, *and of a dairy aspect otherwise*, that was a poor milker. We believe it will be highly conducive to the success of the dairyman to regard this mark as one of chief importance, as it is also to the breeder in his use of animals.

But in laying much stress upon the escutcheon we would not advocate the following of Guenon's classification in all its minute details. It seems sufficient, as far as our own observation suggests, that there should be much of it, symmetrically disposed, and showing on the two rearward lobes of the udder an oblong mirror of large extent.

Were we asked to present a scale of points for the guidance in breeding the model Ayrshire cow, it would take the following form: —

Udder. Capacious, broadly attached, extending far forward and back, closely held to the body, the under surface broad and flat, no clefts, no hollows.

Teats. Shortish, cylindrical, of good texture, and set evenly at considerable distance apart.

Milk Veins. Large, tortuous, disappearing into the abdomen by an ample orifice.

Escutcheon. Large, extending well upwards and on thighs, mirror marks large.

Belly. Ample, broad, deep, and well held up.

Head. Shortish, forehead wide; well set on neck.

Nose. Fine between muzzle and eyes.

Muzzle. Open and moderately large.

Ears. Thin and orange colored.

Horns. Widely set on and of moderate size.

Neck. Of medium length, and straight from head to the top of the shoulders, free from loose skin, fine at its junction with head, and tapering by the symmetrical enlarging of the muscles towards the shoulders.

Shoulders. Thin.

Brisket. Light.

Fore-quarter. Appearing thin in front from the contrast with the hind-quarters and belly, but of sufficient thickness to ensure vigor.

Back. Moderately long and straight.

Spine. Well defined at shoulders, loose jointed, yet level.

Short Ribs. Slightly arched, the concavity in the flank at their extremities responding quickly to digestive condition.

Body. Deep at the flanks and in rib, broad.

Pelvis. Long, broad, and straight.

Buttocks. Neither pointed, nor round and fleshy.

Hook Bones. Wide apart, not overlaid with fat.

Thighs. Thin and broad.

Tail. Long and slender, tufted, set on level with back.

Legs. Short, the bones fine, flat, and the joints firm.

Skin. Soft, moderately thick, loose and elastic.

Hair. Soft and woolly, close.

Temper. Quiet and docile, with nerves suppressed rather than active.

Color. To please the breeder, but not roan.

It is not merely necessary that the Ayrshire should conform outwardly to this type: somewhat more is needed, as will be seen from the following considerations.

Three systems, to which all the organs are directly or indirectly subsidiary, are united in the plan upon which the cow is formed. There is the nutritive system, composed of stomach, intestines, liver, pancreas, glands, and vessels by which food is elaborated, effete matter removed, the blood manufactured, and the whole organization nourished. This is the commissariat. Then there is the nervous system, which co-ordinates all the organs and functions, and enables the animal to entertain relations with the world around it, directing it what to avoid and what to approach, and without which so much complexity of structure as finds place in her organization would be constantly at fault. There is, again, the reproductive system, by which a succession of animals is secured, and the hold of the race on earth assured.[1]

The breed of cow that we should desire must have these systems, each in health and order. Each and all should be developed, not alone to pursue their relations fitly, and to serve their own specific uses, but all developed in the direction to render the animal adapted to serve a particular use, viz. in the dairy cow, the greatest possible utilization of food in the production of a good quality of milk.

The cow in nature lives to one end, the keeping

---

[1] These remarks are suggested by, and adapted from, Dr. Clarke's Sex in Education.

alive the race, as plants have all their energies concentrated to produce seed. The domesticated cow lives to continue her race and to nourish human beings, so that every support must be given to whatsoever will tend to develop her whole organism into the form and activity that conduces to this double service.

Given the proper organs, perfect in form and adaptation to ends, the cow may be a poor thing, if there is sluggish movement throughout her whole organization. To produce a quart of milk an hour, or half or quarter this quantity, signifies activity of organs. This will appear when we consider the waste and renewal of parts that is a phenomenon of life.

Carpenter, in his Physiology, says the whole structure originates in a single cell; that this cell gives rise to others analogous to itself, and these again to many future generations; and that all the varied tissues of the animal are developed from cells. As fast as one cell is destroyed another is generated. The death of one is followed instantly by the birth of its successor. This continual process of cellular death and birth, the income and outgo of cells, that follow each other like the waves of the sea, each different yet each the same, is metamorphosis of tissue.

Bichat has defined life to be organization in action. The most productive cow, as the most productive man in mental or physical labor, lives an intense life. Life, in the sense of motion, — birth and death of cells in the organism, — is lived doubly when the activity is of double measure.

In choosing a milch cow, it may be inferred from this, her quality will not always find expression in outward form. It is the outward form, and the engine within the form, and the *power* of this engine, that is to be looked to. It may occur that animals of two breeds may be found of one form, but that one breed shall carry a quicker life than the other. This quickened life we conceive to be one of the marked characteristics of the Ayrshires.

It is the animal that unites this vitality in a form that favors most economical production, with parts adjusted in symmetrical relations and proportions, that constitutes the perfect cow. When uses are satisfied, then the possession of artistic excellence, such as shall make her a thing of beauty, to the connoisseur as well as the plain farmer, is desired.

THE AYRSHIRE BULL.

The points of the Ayrshire bull should be in many respects those of the cow, but modified by sex. His head should be broad, the muzzle good-sized, the nose small, the under jaw short, and the throat nearly free from hanging folds; his horns should have size at the root, as indicating vigor; his ears thin, and of a golden color within; his eye mild but intelligent, the expression, a masculine vigor, superadded to the feminine type.

His neck should be not too short, but arched, as giving that style which is so attractive. The muscles strong and large, as being masculine; symmetrical in their development, and should not originate from too limited attachments. By this we do not mean a heavy neck, but large in those muscles alone which lie on the sides, well up, and which are so peculiarly masculine in their character as not to be unduly transmitted to female progeny.

The shoulders should be close to the body and thin, the back broad, the spine not as well defined at the shoulders as in the cow, nor the hips quite as broad. The broader the hip, however, the better; yet breadth of hip being more of a female characteristic, this point means more in a slight variation in the bull than in the cow. The pelvis should be long, broad, and straight, and the tail set on level with the back, and without a notch at its insertion. The body should be well-ribbed, deep, and of good length, especially deep in the flanks. A hollow space behind the shoulders is extremely objectionable, as indicating deficiency of constitution. The limbs should be short, fine-boned, flat-boned, and firm-jointed. A curve in the hind leg from the hock to the hoof in front is very objectionable.

In the thigh and the hinder parts must we especially look for the indications of quality for the transmission of milk-yielding shapes. The thighs should be thin and flat, and so far apart as to give great space between. Watch the animal in his walk when going from you. Should his legs cross, reject him. Value him, if in standing his hocks are well apart. The dairy bull should transmit to his female offspring the space for the lodgment of the udder, for this is the key to breeding for milk. Look for the escutcheon, for it is a good feature; and if there is vascularity to be seen here, prominently indicated by the skin and the veins, give heed to it. We also like to see the presence of teats, and the better placed these are the more valuable the indication.

The dairy bull must conform to the type of the ideal dairy cow. The points which indicate digestive ability and space for the lodgment of the udder must be strongly characterized. Hardly less important is that feminine appearance joined on to a masculine vigor, which is shown in part by a noble serenity of expression. '

The dairy bull must be especially free from certain defects and blemishes, such as the fleshy buttock and rounded ham which is followed by roundness of thigh. It is a beefy mark. He must not be too heavy forward. Not that we would have the strongly developed wedge shape of the cow, but we would have those points which we value so highly in the rearmost-half of the cow strongly developed in the bull.

These characters in the Ayrshire bull must be united in that peculiarity of shape which is indicative of the breed, and which is so difficult to put into words. There should be style, the long, slim tail and bushy switch, the harmony of proportion, and the moulding of the lines of his contour, and all else which constitutes beauty.

The grazier and the dairyman have each for a long time sought to improve his stock, the one for meat, the other for milk. The union of the two in an animal in equal excellence as though they existed in perfection separately, we believe is never realized; so that whenever it is desirable to have milk or meat in large amount, at very low cost, it is better to cultivate these qualities in separate breeds, and encourage in each that particular development conducive to the quality desired in them.

No one will hesitate to affirm that the meat breeds have arrived at a greater excellence in the line of development for meat than have the dairy breeds in the line of development for milk. The ideal of the breeder has been nearer realized. Various causes have contributed to this. It is only in place here to refer to one as having direct relation to the breeding of Ayrshires, and as directing us to a knowledge of what we should desire in a bull of the dairy breed.

If we refer to the Short-horn, the Hereford, the Galloway, and many other races that are acknowledged meat breeds, and observe the type of the male and the female, we find it essentially the same for either sex. In the cow and in the bull there is an approach to one form, modified only by such differences as attend upon sex.

With many dairy breeds, perhaps with all, many persons refuse to accept the principle that directs the grazier, and conceive the cow should be of one type, the bull of quite different type. How frequently it happens at our agricultural fairs that we observe a fine dairy herd of cows possessing the forms that go with the possession of dairy qualities in a high degree, headed by a bull whose outlines are those of a meat bread! In the one sex the outline is that of the keystone of an arch, in the other the brick form pediment. Here is being used two typical forms of distinct functions, to effect one form in the female line and another in the male line.

Consider the comparative ease with which the grazier obtains a bull satisfactory to him with the difficulty of the dairyman to realize his idea in the bull!

The meat breeds were early taken in hand by master breeders and were greatly improved. To these men we owe most of the maxims and current knowledge of the principles of breeding. One of these maxims is "Like begets like." The grazier uses *like* in male and female, and gets *like*. The dairyman often uses one form of animal in the female, the form typical for the females of the breed, a form in the male animal of a dissimilar type, and expects that *unlike* in each generation will get a *uniform* progeny.

We know persons who own Jersey and Ayrshire stock, who have their conception of the male animal so much fashioned by the model to which a meat breed should correspond, that they seek far and wide to place at the head of their herd a bull conforming to this foreign mould, although they esteem less valuable the female progeny that may conform to it.

II.

HISTORY.

In those cases where documentary evidence is obscure, and but few notices concerning the origin of a breed are attainable, it seems the more philosophical to first study the condition of the country and the surroundings from which the breed was evolved.

Our scheme concerns itself first with the earlier records. After presenting in general terms some aspects of the Scotland of the past, we shall next call your attention to the cattle known as the White Forest Breed. A sketch of the county of Ayr of the past and present will naturally precede the division we have devoted to the presentation of the documentary evidence concerning the origin of the Ayrshire breed; and the history of their origin, as founded on the preceding chapters, will follow next in course.

SCOTLAND AND ITS PAST.

HORNED cattle are said to be indigenous to Scotland. From prehistoric research, Wilson[1] states, on evidence no doubt satisfactory, that in these early times "vast herds of wild cattle of gigantic proportions and fierce aspect roamed through the chace."

The earliest historical notice of British cattle are in the "Commentaries" of Cæsar, in which he mentions their abundance, and that the food of the inhabitants was milk and flesh, to the neglect of tillage; and Strabo[2] praises the bountiful supply of milk, but denies to them the art of making cheese.

Darwin[3] states that *Bos primigenius* existed as a wild animal at this time, and that *Bos longifrons* was domesticated in England during the Roman period, and supplied food to the Roman legionaries.[4]

At this early period, the savage time, so to speak, the same cattle seem to have been found more or less on both sides of the border; and in considering the wild cattle of Scotland, it will be useful to review in some measure the cattle of England, and the state of the country in those days. Fitz Stephen,[2]

---

[1] Prehistoric Annals of Scotland.
[2] James Wilson, in Enc. Brit. xiv, 214.
[3] Animals and Plants under Domestication, N. Y. 1868, 1, 104.
[4] British Pleistocene Mammalia. Dawkins and Sandford, p. xv.

who lived in the twelfth century, speaks of the *Uri Sylvestris*, which in his time inhabited great forests in the neighborhood of London; and in the fourteenth century King Robert Bruce was nearly slain by a wild bull which attacked him in the great Caledon Wood.[5]

Boethius,[6] who was born in 1470, and John Leslie, Bishop of Ross, who wrote in 1598,[7] state that the wild cattle of Scotland were white with a thick mane; and Leslie expressly states were wild and savage, and formerly abounded in the Caledonian Woods, but now were confined to the region about Sterling, Cumbarnauld, and Kincardine.

At this period civilization had made some progress in the country, the Lowlands at least; and food, judging from fragments of history, was bountiful and cheap. In 1290 the monasteries of Teviotdale had much pasture land, and the minute and careful arrangement of their mountain pastures, of the folds for their sheep, and the byres for their cattle, and the lodges or temporary dwellings for their attendants, show that they paid the greatest attention to this part of their extensive farming.[8] Again in 1300, from excerpts[9] from the reign of Alexander III, we have it stated that the fields, the mountain pastures, and the forests were amply stocked with cows, sheep, and large herds of swine; and even more minutely,[10]

---

[5] Cosmography and Description of Albion, quoted in Enc. Brit. xiv, 214.
[6] Annals and Mag. of Nat. Hist., vol. ii, 1839, p. 281; vol. iv, 1849, p. 424.
[7] Low's Animals, p. 234.
[8] Innes' Scotland in the Middle Ages, p. 147.
[9] Tyler's Hist. Scotland, ii, 218.
[10] Tyler, *op. cit.* p. 221.

that in the more cultivated districts cows were kept in the proportion of ten to every plough, but in the wilder part of the country the number was infinitely greater.

It is to be supposed that with such numerous cattle, and with such indications of the amount of pasture land, some reference would be made to the dairy; and, indeed, from these same excerpts,[11] we find that much cheese was manufactured on the royal demesne throughout Scotland, and as Tyler remarks, "It is equally certain that its proper management and economy was not neglected by the clergy or the barons."

This period seems to have been one of plenty even bordering on luxury, and it is most probable some attention was given at this time to the improvement of the domestic breeds. We know from the Cartularies of Melrose and Kelso[12] that in the fourteenth century, many of the nobles had breeding studs on their estates, and from Rotuli Scotiae we learn that Lord Douglas brings ten "great horses" into Scotland, July 1, 1352.

King Alexander, who ruled during the last half of the thirteenth century, showed an interest in husbandry, and caused a great breadth of land to be brought under the plough; and according to that quaint chronicler in rhyme, Andrew Wyntoun, "Corn he gart be aboundant."[13]

These good old times passed away (all Golden Ages

---

[11] Tyler, *op. cit.* p. 221. [12] Tyler's *op. cit.* ii, 218.
[13] Cronykil of Scotland.

are ancient), and in the history of fightings and treacheries and murders, and the great ones of the land, we lose account of the agriculture. It is only to be gathered that the Age of Iron succeeded this halcyon age of the thirteenth century.

In 1570 Ortelius[14] describes the cattle of the county of Carrick as being of large size, with tender, sweet, and juicy flesh; and our references are thus brought to the earliest mention of the cattle occupying the county of Ayrshire.

It is thus seen that cattle are natives of the isle. Their first appearance is neither recorded by history nor by tradition, and their remains in cairn and cavern place their antiquity beyond our written records. Thus, in a sense, they are autocthenes, or the product of the soil. They existed in a wild state as late as A. D. 1200 in the neighborhood of London, and in 1600 occupied, in a state of freedom, a circumscribed locality in Scotland. During this whole period domestic races existed in their close vicinity, and the economy of the dairy in A. D. 1300 seemed to be well understood.

Our records, it will be perceived, refer to the Lowlands of Scotland. These were conquered by Agricola, and his conquest secured by a chain of forts connecting the Firths of Forth and Clyde, A. D. 85.[15] But so courageous and indomitable were the barbarians, that under the Emperor Hadrian, about A. D. 120, a fortified rampart was constructed from

---

[14] Theatrum orbis Terrarum.   [15] Enc. Brit. xix, 741, 743.

the Tyne to the Solway. In 207 the country north of the Clyde was savage and uncleared, and the fierce inhabitants in A. D. 446 are described as bearing all the stamp of barbarian life.

Upon the abandonment of Britain by the Romans in 446, the five tribes inhabiting the region about these fortifications became independent, and under a union formed a new kingdom termed Regnum Cumbrense, or more frequently the Kingdom of Strathclyde. It appears to have included the present Liddesdale, Teviotdale, Dumfriesshire, Galloway, Ayrshire, Renfrew, Strathclyde, the midland and western parts of Stirlingshire, and the largest portion of Dumbartonshire. The tribes which formed this community led a pastoral life, living on the milk of their flocks and the produce of the chase. They were a race not of different descent from the ruder tribes of the north, but of improved civilization.[16]

This was the region in which civilization first obtained a foothold, and where the labors of agriculture must first have taken the place of the uncertainties of the chase. The only early record of dairy products is from within this district,[17] as well as the breeding of horses. As an evidence of culture existing here at an early date, it may be well to state that the immense Abbey of Kelso was begun in 1128, and the beautiful Melrose Abbey in 1136.[17]

Until the middle of the eighteenth century there was scarcely a good road in Scotland.[18] In 1678 an

---

[16] Enc. Brit. xix, 741, 743. [17] Black's Picturesque Guide to Scotland.
[18] Enc. Brit. xix, 807.

agreement was made to run a coach with six horses between Edinburgh and Glasgow, forty-four miles, the double journey to be made in six days, and the common carrier occupied a fortnight in journeying to and from Selkirk and Edinburgh, a distance of thirty-eight miles.[19] If such were the roads in the more cultivated districts, communication must have been exceeding difficult in the Highlands. There, agriculture was neglected, the circumstances of the climate, soil, and disposition of the people were unfavorable; and in 1787 even, the imperfect infant state of the agriculture[20] may be inferred, by the coast inhabitants or those of the isles obtaining a greater part of their subsistence by fishing, while the more inland clans depended chiefly on their cattle and flocks. In 1714, in the island of Stroma, in Caithness, there was but one small plough.[21] In 1799 the roads in many places did not pass by a single village, house, hut, or inhabitant, for fifteen or twenty miles.[22]

The influence of these two states of affairs in the Highlands and Lowlands are seen in the cattle. In the more nomadic state of the Highlands we find but one style of cattle, the Highlanders,[23] — animals of strong individuality, varying among themselves according to the luxuriance of the pastures and the effect of climate, from the diminutive Shetland, the ordinary West Highlander, North Highlander, and the Runts, to the well-formed Argyleshire.

---

[19] Enc. Brit. xix, 807.
[20] Prize Essays H. Soc. 1st ser. vol. i, p. 129.
[21] Ibid. vol. i, p. 128.
[22] Ibid., i, cxiii.
[23] Youatt and Martin on Cattle, passim.

In the Lowlands, on the contrary, on account of the varied wants superinduced by civilization, we have a multiplicity of breeds, each best suited to the exigencies of their locality, either at the present or some past period. It is sufficient to mention the Galloway and the Ayrshire. Where the Lowland characteristics and civilization invade the geographical Highlands, we find the shapes and uses of the cattle modified, as in the Aberdeenshire and Angus breeds.

The most cultivated breed, the highest product of civilization, occurs in that locality where the civilization is the more ancient. We refer to Ayrshire and the Ayrshires.

Without other information, it is unreasonable to suppose, with the evidence of wild cattle being domesticated in England, that the present cattle of Scotland are derived entirely from importations, rather than founded on the original stock, modified, in what way you please, by successive crossings or systems of breeding.

We are accordingly led to examine into what has been chronicled of the wild cattle of Scotland, and to investigate what claims they have to be considered in the light of a foundation upon which the improved breeds have been builded.

THE WILD CATTLE OF SCOTLAND, OR WHITE FOREST BREED.

According to our best authorities, two forms of the ox tribe, the genus *Bos*, existed in Scotland at an early period, *Bos primigenius* and *B. longifrons* of Owen. The former was of large size, and according to all accounts the color was black; it had white horns with long black points, the hide was covered with hair shorter and smoother than in the tame ox, but on the forehead long and curly. From the skeletons preserved in our museums, the length of this gigantic ox must have been from eleven and one half to twelve feet, and the height of the shoulders about six or six and one half feet.[1] Darwin remarks that the Pembroke race in England closely resembles this ox in essential structure, and that the cattle at present existing in the Chillingham Park are degenerate descendants of this breed.[2] *Bos longifrons*, on the contrary, is described as a distinct species, of small size, short body, and fine legs. It was domesticated in England during the Roman period.[3] Professor Owen thinks it probable that the Welsh and Highland cattle were descended from this species.[4]

---

[1] Nilsson, Annals and Mag. of Nat. Hist. 1849, iv, 258.
[2] Animals and Plants under Domestication, i, 103.
[3] British Pleistocene Mammalia, p. xv.
[4] Animals and Plants under Dom. i, 104.

WHITE FOREST BREED.

In prehistoric times, a continuous range of enormous forests covered the whole extent of the country, and the gigantic and fierce cattle roaming through the chase[5] fed on the tender branches and buds, the catkins of birch, hazel, sallow, and other species of willow,[6] resembling in this matter of feeding the moose of the Canadian forests. We have reason to suppose that the ancient islanders introduced the rudiments of a pastoral life, while yet living in pits incovered with boughs and skins,[7] yet no evidence leads to the conclusion that the native Britons had domesticated the great oxen of the country, although undoubtedly they formed a source of food.[8] In Switzerland, on the contrary, the lake dwellers had succeeded in taming these formidable brutes.[9]

We have it stated by Darwin, that *Bos primigenius* existed as a wild animal in Cæsar's time.[10] There is a record of white cattle in the tenth century, resembling those in the Scottish parks, existing in Wales, where they were more valued than black cattle.[11] Boethius, in 1526, mentions them as then existing near Stirling. " At this toun began the grit wod of Calidon. This wod of Calidon ran fra Striveling throw Menteith and Stratherne to Atholl and Lochquabir, as Ptolome writtis in his first table. In this wod wes sum time quhit bullis, with crisp and curland mane, like feirs lionis, and thought thay semit meek

---

[5] Prehistoric Scotland, Wilson's.
[6] Nilsson, An. & Mag of Nat. Hist. 1849, iv, 289.
[7] Prehistoric Scotland, 1, 296.
[8] Ibid. i, 31.
[9] Lyell's Antiq. of Man. Phila. 1863, p. 24.
[10] Animals and Plants under Domestication, 1, 104.
[11] Low's Animals, 239.

and tame in the remanent figure of thair bodyis, thay wer mair wild than ony uthir beistis, and had sic hatrent aganis the societe and cumpany of men that thay come nevir in the wodis, nor lesuris quhair thay fand ony feit or haind thairof, any mony dayis eftir, they eit nocht of the herbis that wer twichit or handillit be men. Thir bullis wer sa wild, that thay wer nevir tane but slight and crafty laubour, and sa impacient that eftir thair taking they deit for importable doloure. Alse sone as ony man invadit thir bullis, they ruschit with so terrible preis on him, that they dang him to the eird, takand na feir of houndis, scharp lancis, nor uthir maist penetrive wapinnis." " And thoucht thir bullis were bred in sindry boundis of the Calidon wod, now, be continwal hunting and lust of insolent men, thay are distroyit in all partis of Scotland, and nane of thaim left bot allanerlie in Cumarnauld."[12] In a remarkable document, written about 1570, the writer complains of the aggressions of the king's party in the destruction of the deer in the forest of Cumbernauld, " and the quhit ky and bullis of the said forest, to the gryt destructione of policie and hinder of the commonweill. For that kynd of ky and bullis he bein kepit thir money zeiris in the said forest, and the like was not mantenit in ony vther partis of the Ile of Albion."[13] In 1598, John Leslie, Bishop of Ross, speaks of the wild ox occurring in the woods of Scotland, of a white color,

---

[12] Hector Roscoe, born in 1470. Hist. Scotorum, pub. at Paris, 1526, ed. of 1574, fol. 6, line 63, occurs the passage quoted in An. & Mag. of Nat. Hist. 1839, ii, 281, and Low's Animals, 234.

[13] Illustrations of Scottish History, preserved from manuscripts by Sir John Graham Dalyell, Bart., quoted in Low's Animals, p. 235.

with a thick mane resembling a lion's, and wild and savage. He says that it had formerly abounded in the Sylva Caledonia, but was then only to be found at Stirling, Cumbernauld, and Kincardine.[14] Sandford, in his manuscript history of Cumberland, dated 1675, says around Naworth formerly were " pleasant woods and gardens; ground full of fallow dear fieding on all somer-tyme; brawe venison pasties, and great store of reid dear on the mountains; and white wild cattle, with black ears, only on the moores."[15] We find them referred to by Bewick in 1770, and in 1781 Pennant speaks of them as retaining their white color, but as having lost their manes.[16] Conrad Gesner describes them as " white oxen, maned about the neck like a lion. . . . This beast is so hateful and fearful of mankind, that it will not feed ot that grasse or those hearbes whereof he savoureth a man hath touched — no, not for many days together; and if, by art or policy, they happen to be taken alive, they will die with very sudden grief. If they meet a man, presently they make force at him, fearing neither dogs, spears, nor other weapons."[17]

About 1800 they are spoken of as invariably white, with the ears internally and externally about one third down, red; horns white, tipped with black, and the muzzles black.[18] In 1836 we begin to get more particular descriptions. Color invariably white, muzzle

---

[14] Leslie. De Origine Moribus et Rebus Gestis Scotorum, Rome, 1598, ed. of 1675, 18, quoted in An. & Mag. of Nat. Hist. 1839, ii, 282. Also in Low's Animals, 234.
[15] Jour. R. A. S. 1852, xiii, 219.
[16] Quadrupeds, 16.
[17] 16th Century; quoted from Scherer's Rural Life, p. 627.
[18] Complete Grazier, p. 1.

black, the whole of the inside of the ear, and about one third of the outside, from the tip downward, red. The horns are very fine, white with black tips; and the head and legs are slender and elegant.[19] The Earl of Tankerville, the proprietor of Chillingham Park, describes them in 1839. In form they are beautifully shaped, with short legs, straight back, horns of a very fine texture, as also their skin, so that some of the bulls appear of a cream color.[20] In 1845 Low says that the eyelashes and tips of the horns are black, the muzzle brown, the inside and a portion of the external parts of the ears are reddish-brown, and all the rest of the animal white. The bulls have merely the rudiments of manes, consisting of a ridge of coarse hair upon the neck.[21] In 1852 William Dickinson says that their bodies are pale cream color, the ear-tips red, and the muzzle black.[22] In 1868 Darwin describes them as white, with the inside of the ears reddish-brown, eyes rimmed with black, muzzle brown, hoofs black, and horns white tipped with black.[23] Youatt mentions the existence of a mane on some of the bulls, one and one half or two inches in length.[24]

As a wild race we hear of their occurrence at rare intervals. In the time of Edward the Confessor (1042) we are told by one of the abbots of St. Albans that wild bulls abounded near London,[25] and Fitz-Stephen, writing about 1174, speaks likewise of their occurrence in these woods.[26] In 1760 wild

---

[19] Naturalists' Lib. Jardine. iv, 202.
[20] An. & Mag. of Nat. Hist. 1839, ii, 277.
[21] Low's Animals, 237.
[22] Jour. R. A. S. 1852, xiii, 249.
[23] An. & Pl. under Dom. 107.
[24] Youatt & Martin on Cattle, 12.
[25] An. & Mag. Nat. Hist. 1st ser. iii, 356.
[26] An. & Mag. Nat. Hist. 1849, iv, 423.

white cattle were just extinct in the central Highlands.[27] In 1598 their occurrence in Scotland was confined to a few localities.[28] We are thus particular in tracing the accounts of this breed, as Wilson maintains that no sufficient evidence has ever been brought forward to prove that these cattle are entitled to the character of an aboriginal breed.[29] Of the remnants of this ancient race we have two herds, at least, existing at the present time, and records of others whose extinction has been comparatively recent. The general descriptions of white with colored ears apply to all, yet each herd has had its distinctive features, and we find evidence of a constant tendency to variation, only repressed by a rigorous selection.

Chillingham Castle, the seat of the Earl of Tankerville, is situated in Northumberland County, England, and formerly occupied one end of the Caledonian Forest, which in former times extended from sea to sea. The wild cattle have been preserved in this park with care, and kept free from intermixture with other breeds. They have been extensively inbred from necessity, "and are accordingly much subject to rash, a complaint common to animals bred in and in." According to Denholm, they were exterminated in 1760. "Here (Cadzow Castle) so late as the year 1760 were a few of those white cattle with black or brown ears and muzzles, once so common in Scotland. Their shyness and ferocity of temper rendered

---

[27] Trans. H. & Ag. Soc. 4th series, v, 294. [28] Low's Animals, 234.
[29] Enc. Brit. xiv, 214.

them troublesome and of little use; they were therefore exterminated in the year above mentioned."[30]

We find it recorded that the stock at Chillingham was at one time left without a bull, from accident and sterility. Fortunately one of the cows had a bull calf, and the stock was preserved.[31] In color, they are invariably white,[32] or white[33] or pale cream color,[34] or creamy white,[35] or white and cream color.[36] Their horns are white tipped with black; their muzzle black[37] or brown;[38] their eyelashes black;[39] their eyes rimmed with black.[40] Their ears inwardly and about one third externally, red,[41] reddish-brown,[42] or red or brown.[43] Their necks have rudimentary manes,[44] or oftentimes a mane from one and a half to two inches long,[45] or no manes but coarse hair.[46] Their heads slender,[47] backs straight. Legs short[48] and slender,[49] and the hoofs black.[50]

In 1675, as we have seen, they are described with black ears.[51] In 1770, according to Bewick, some

---

[30] The History of the City of Glasgow, etc., by James Denholm. Glasgow, 1798. p. 252.
[31] Earl of Tankerville, Annals and Mag. of Nat. Hist. 1839, ii, 284. Nat. Lib. Jardine, iv, 207, note.
[32] Nat. Lib., Jardine, iv, 202, note.
[33] Darwin, An. & Pl., under Dom. 1, 107.
[34] Hindmarsh, An. & Mag. Nat. Hist. 1839, ii, 279. Dickinson, Jour. R. A. S. Eng. 1852, 249.
[35] Capt. Davy, Milk Journal, Oct. 2, 1871, 225.
[36] Earl of Tankerville, Annals of Nat. Hist. 1839, ii, 277.
[37] Dickinson, Nat. Lib.. Capt. Davy, op. cit.
[38] Low, Darwin, Earl of Tankerville, op. cit.
[39] Low, Hindmarsh, op. cit.
[40] Hindmarsh, Darwin, op. cit.
[41] Dickinson, Nat. Lib., op. cit.
[42] Low, Darwin, op. cit.
[43] Earl of Tankerville, Annals of Nat. Hist. 1839, ii, 277.
[44] Low's Animals. p. 237.
[45] Youatt and Martin on Cattle, p. 12.
[46] Earl of Tankerville, An. of Nat. Hist. 1839, ii, 277.
[47] Earl of Tankerville, An. of Nat. Hist. 1839, ii, 284.
[48] Earl of Tankerville, An. of Nat. Hist. 1839, ii, 277.
[49] Nat. Lib., Jardine, iv, 202, note.
[50] Darwin, An. & Pl. under Dom. 1, 107.
[51] Jour. R. A. S. 1852, xiii, 249.

calves appeared with black ears, but these were destroyed, and black ears had not since reappeared.[52] Since 1855 about a dozen calves have been born with brown or blue spots on their cheeks or necks, but these, with any other defective animals, were immediately destroyed,[52] and Low speaks of the tendency of the young to be altogether black or altogether white, or to have black ears.[53] In Keux's " Natural History," published probably in the earlier part of the present century, these cattle are said to have lost their manes, but to have retained their color and fierceness; to be of a middle size, long legged, with black muzzles and ears, and their horns to be fine and to have a bold and elegant bend. The keeper of those at Chillingham said that the weight of the ox was thirty-eight stone, of the cow twenty-eight. It would thus seem as if Keux spoke from personal observation.

Dr. Knox remarks that the wild white bull of Scotland, instead of having large horns like the fossil breed, has either comparatively short horns, or none at all; and when present they follow precisely the direction observed in those of the surrounding domestic breeds. He also says that when calves are taken from the cow and brought up with the domestic cattle of the neighboring farms, they grow up quite gentle, and precisely as other cattle. When the young are born with red or black spots, or without tails, or very short ones, they are uniformly de-

---

[52] Darwin, An. & Pl. under Dom. i, 107.
[53] Low's Animals, 238.

stroyed, the noble proprietor considering the white color to be essential to their purity.[54] Mr. Cole, the park-keeper for more than forty years, says they have no mane, but curly hair on their neck and head; more so in winter, when the hair is long.[55] Cully says their color is invariably of a creamy white; muzzle black; the whole of the inside of the ear and about one third of the outside, from the tip downwards, red; horns white, with black tips, very fine, and bent upwards; some of the bulls have a thin, upright mane, about an inch and a half or two inches long.[56]

The Hamilton Park cattle are often referred to as the cattle of the Chase of Cadzow, after the castle of that name, the former seat of the dukes of Hamilton. Cadzow Castle occupies a site on the banks of the Avon in Lanarkshire, at one extremity of the ancient Caledonian Wood. Aiton, in 1814, describes these cattle as uniformly of a creamy white color, their muzzles and the greater part of their ears black or brown, and some with a few black spots on their sides. A few are without horns, but the greater number have handsome white ones, with black tips bent like a new moon. Some of the bulls have a sort of mane, four or five inches long. The cattle at Hamilton and Ardrossan are not so fierce and savage as their ancestors, but at Auchencruive they still retain much of their natural ferocity. Their backs are high and not so straight as could be wished; their chest

---

[54] Jour. of Ag. ix, 372, 376. [55] Vasey on the Ox Tribe, p. 149.
[56] Vasey, *op. cit.* p. 143.

is deep but narrow; and they have much the appearance of the ill-fed native breed of the cattle of Ayrshire, Lanarkshire, etc., about fifty years ago.[57] In 1845 Low describes them as with the females generally polled,[58] and in 1870 the bulls are credited with black-tipped horns.[59] Their color is given as dun white,[60] or dingy white,[61] their muzzles and hoofs black,[62] as also the inside of the ears,[62] and the tongue.[63] In the "Naturalists' Library" we find it stated that their bodies are thick and short, their limbs stouter than the Chillingham breed, and their heads much rounder, the inside of their mouths either black or spotted with black, and the fore part of their legs, from the knee downward, mottled with black.[64] At one time but thirteen remained alive, the survivors of the cattle-plague of the few years previous. The bulls looked as if they might fatten to eight hundred or eight hundred and fifty pounds. They had light hind-quarters, but were heavy and deep in front; all had black muzzles, black ears, and the older beasts black tips to their horns.[65] We were told that some years ago the herd numbered eighty or ninety, but all fell victims to the cattle-plague except thirteen, of which eleven altogether escaped and two recovered. When the plague attacked them, they were driven individually between

---

[57] Sinclair's Scotland, iii. 44.
[58] Low's Animals, 236.
[59] Gard. Chron. and Ag. Gaz., Aug. 6, 1870.
[60] Low. Nat Lib., *op. cit.*
[61] Dickinson, Jour. R. A. S., of Eng., 1852, 249.
[62] Low, Nat. Lib., *op. cit.*
[63] Low's Animals, 236.
[64] Nat. Lib., Jardine, iv, 202, note.
[65] Gard. Chron. and Ag. Gaz., Aug. 6, 1870.

gradually approaching fences, leading to a large and strong wagon sunk to the ground level, and so captured and taken to separate abodes, where they were confined until all risk was passed. They have now (in 1870) increased to thirty-seven.[66]

Dr. Knox says of these animals that they differ a good deal in form from those of the Chillingham Park. The markings also are different; but still there is a strong tendency in the young cattle to cast calves which are said to be "off the markings," and to be either entirely black or entirely white, or black and white, but never red or brown.[67]

We have mention of some having been kept at Ardrossan and Auchencruive, but no further particulars, except that those at the latter place were very fierce.[68] They were also kept at Bishop-Auckland in 1635.[69]

The cattle preserved at Drumlanrig, the seat of the Duke of Queensberry, are said by Darwin to have become extinct in 1780, and are described as with their ears, muzzle, and orbits of the eyes black.[70] Pennant, writing in 1781, speaks of them as still existing, having lost their manes, but of a white color.[71] Dickinson states that two cows and a bull were living in 1821, but the bull and one of the cows died that year. He describes them as dun or rather flea-bitten white, polled, with black muzzles and ear-tips,

---

[66] Gard. Chron. and Ag. Gaz., Aug. 6, 1870.
[67] Jour. of Ag. ix, 376.
[68] Sinclair's Scotland, iii, 44.
[69] An. Nat. Hist. vol. iii, ser. 1, p. 241.
[70] Darwin, An. and Pl. under Dom. i, 107.
[71] Quadrupeds, 16.

with spotted legs.[72] Low says they were destroyed many years ago by order of the late Duke of Queensberry.

The cattle at Gisburne Park, in Craven, County of Yorkshire, England, the seat of Lord Ribbesdale, are mentioned as late as 1852, as being pure white with brown or red ears and noses.[73] Low speaks of their being polled,[74] and Bewick describes them as perfectly white except the inside of their ears, which are brown. They are without horns, very strong boned but not high.[75] He also states, as Darwin quotes, that they are sometimes without dark muzzles.[76] They are said to have been originally brought from Whalley Abbey, in Lancashire, upon its dissolution in 1542.[77]

The herd at Burton Constable, also in Yorkshire, situated in the District of Holderness, all perished in the middle of the last century of an epidemic disorder. They were of large size, and had the ears, muzzle, and tip of the tail, black.[78]

From Garner's "National History of Staffordshire," we learn that the wild ox formerly roamed over Needwood Forest, and in the thirteenth century William de Farrarus caused the park of Chartley to be separated from the forest; and the turf of this extensive enclosure still remains almost in its primitive state. Here a herd of wild cattle has been preserved

---

[72] Dickinson, Jour. R. A. S. of Eng., 1852, 249.
[73] Dickinson, *op. cit.*
[74] Low's Animals, 238.
[75] Bewick's Quadrupeds, 8th edit. 39, note.
[76] An. and Pl. under Dom. i, 108.
[77] Bewick's *op. cit.*
[78] Low's Animals, 238.

down to the present day, and they retain their wild characteristics like those at Chillingham. They are cream colored, with black muzzles and ears; their fine, sharp horns are also tipped with black. They are not easily approached, but are harmless unless molested.[79] Low adds that they frequently tend to become entirely black, and that they are of a larger size than those at Chillingham.[80]

Wild cattle, says Low, have been or are yet preserved at Wollaton in Nottinghamshire and at Limehall in Cheshire, England,[81] and Bewick states that the ears and nose of all of them are black.[82]

These cattle, in the possession of ancestral families, and maintained and protected in parks, undoubtedly as a family pride, have with difficulty been preserved through the epidemics and casualties of a few centuries. Yet, despite the human care and the rigorous weeding out of blemishes, we can see they were unable to retain in their color or form much more than a resemblance. In the Chillingham cattle the muzzle is described as black or brown, the ears inwardly, and in part externally, red, reddish-brown, and red or brown; their manes either short, or rudimentary, or not existing. We find black ears and blemishes occurring at different times. In the Hamilton herd we find them generally with horns at an early date, but afterwards the females usually polled. Black spots on sides and legs are noticed. They are described as possessing manes of from four

---

[79] Vasey on the Ox Tribe, p. 140.
[80] Low's Animals, 238.
[81] Low's Animals, p. 238.
[82] Bewick's Quadrupeds, 8th edit. 39, note.

to five inches long, especially some bulls. Their limbs have become stouter and their heads shorter than the Chillingham breed at the other end of the ancient wood. Those at Drumlanrig have become polled, presumably in both sexes. At Gisburne Park, they are not only hornless, but only the inside of their ears are colored, and occasionally they lose their dark muzzle. At Burton Constable, among their fertile pastures, we see an increase of size, the effect of the abundance of the feed; and the end of their tails have become black. In Staffordshire, we observe the tendency to become entirely black.

When even selection finds it so difficult to preserve the uniformity of the same herd for successive years, and fails even more glaringly when applied to different herds under varied circumstances, we can hardly be justified in rejecting these white cattle, as the primitive or foundation stock of existing breeds of that county on account of their color alone.

The wild state seems peculiarly favorable to uniformity of coloring, as the causes which have operated to produce the result on a few, act likewise upon all, and are constant in their action. Any deviations from the markings appear to become absorbed in the multitude, so as to have little opportunity for preservation. In civilization, on the contrary, we have the element of human will, a highly complex and variable possession, which interrupts the apparent harmony of uncultured nature by rendering new combinations possible and probable.

That a slight interference with a natural state will

produce variability of coloring, is well shown in an account of the cattle of Paraguay, by Azara, wherein it is stated that the wild cattle are always a reddish pard color, and thus differ in color from the domesticated breeds.[83] When it is considered how little tameness is called domestication in these regions, it is realized upon what obscure causes the fact of color must depend. Even in our most ancient breeds we find variations of color, as in the Highland, Galloway, and Devon.[84] The strongest single argument in favor of these white cattle being the forefathers of our present stock, is in the occasional cases of reversion, which occur in many of the breeds, and oftener in those whose connection with the wild breed seems probable. In the West Highland breed, usually black, the white color and the ear markings in many cases return.[85] In the Ayrshire cow we have record of two cases of reversion, to white with red ears; and we can remark, after a most careful examination of Ayrshire cattle, that we have never seen white ears, or ears the tips of which were other than red, brown, or black. In shape we have the differences inherent to locality. Mountain breeds are apt to be lighter in their hindquarters than breeds occupying a plain, as we are told by Low,[86] and it is obvious to any observer that semi-domesticated breeds are lighter in the flanks and loins than those breeds which have been subjected to systematic breeding. In the Ayr-

---

[83] Nat. Hist. of the Quadrupeds of Paraguay, Edinb., 1838, 73.
[84] Low, *passim.*
[85] Low's Animals, 301. [86] Low's Animals, 305.

shire breed, we find the medium horn, often the direction of the curve with the frequent black tip, pointing to the wild breed, as also the white face, or starred forehead, and the "rigged" back occasionally or frequently recurring, to direct our attention to the transition cattle between the original stock, and the recorded results of breeding, coeval with the advanced interest in agricultural pursuits at or about 1800.

These cattle in their present state are easily and readily tamed, and crosses with common stock are occasionally noted. Such with the forest bull are said by Bewick to invariably take the color of the father and to retain some of the fierceness.[87] One recorded instance of the crossing of a cow of the white breed by a common bull, gives the color of the progeny as after the forest pattern, but with mottled legs.[88] Another, between the white bull of the Hamilton herd and a Shetland cow, produced a very good-looking polled ox, "nearly quite black," and greatly superior in weight to the Shetland cow.[89]

When we consider the small number of these cattle, and the length of time they have been preserved, and how narrowly they have escaped utter extinction, it is difficult to suppose that they have been retained in their purity; still less when we consider the disturbances of the times, the number of cattle grazing continually in their vicinity, and the striking resemblance which is often shown to them by cattle of

---

[87] Bewick's Quadrupeds, 41, note.
[88] Hindmarsh, Ann. and Mag. of Nat. Hist. 1839, ii, 280.
[89] Dr. Knox, Jour. of Ag. ix, 369.

other breeds. According to Low, individuals were to be met with in 1845, in the county of Pembroke, in no ways distinguishable from the wild cattle of the Parks,[90] and Aiton speaks of their resemblance to the common cattle of 1750. We have ourselves seen in America, cattle which were pure white with red ears, and even polled.

The only explanation we can offer for the variations between the herds of forest cattle, and the tendency towards variation, which seems from our account to have been ever strong, is that these, as well as the domestic cattle of this region, are offshoots from the same original stock, the wild ox of the past, but that those races we call domesticated, as the Ayrshire, the Angus, the Galloway, the Highland, and others, have been influenced to a greater extent by the arts of civilization, the conscious or unconscious breeding for certain uses, and the effects of crossing, than these inhabitants of the parks.

On this view the White Forest Breed is a wild animal, a descendant, with now and then a bar sinister, of the wild breed; and the domesticated races of the country are likewise their descendants, but with an ancestry hopelessly confused and intermixed by outside crosses and influences.

---

[90] Animals, 296.

THE COUNTY OF AYRSHIRE.

In the south of Scotland, on the western coast, lies the County of Ayrshire. The outline of its boundaries encloses a crescent-shaped area, with the concavity towards the sea, — its length about eighty miles, and its breadth varying from a few miles at the extremities to about twenty-eight miles in the centre, it contains 1,149 square miles, or 735,262 acres of surface.[1] Generally low adjoining the sea, the land rises by easy slopes and wavy undulations, to a ridge of high or hilly country, in part almost mountainous, which forms its eastern boundaries. No portion can be termed level, for numerous swells or rounded hills give variety to the landscape. As the slope of the land is generally westerly, towards the shore, or the valleys of the streams flowing thither, it follows that the principal exposure of the arable land is westerly and southerly, a fact which is of importance as explaining in part the moderation of the climate. The country is well watered by numerous streams, which, rising among the eastern hills, find their way in a tortuous course to the sea.

---

[1] Jour. R. A. S. of Eng. 1866, p. 426.

Ayrshire is probably the most densely-wooded county in Scotland, although most of the woodland was planted towards the close of the last century and beginning of this. The growth is chiefly of larch and Scotch fir, but generally having hardwood trees intermixed, — beech, ash, and elm predominating.[2] More than one half of the country may be classed as unimproved, being occupied by hills, moors, mosses, and lochs.[3]

Historically and statistically the county is divided into three districts, from north to south. Cunningham comprehends the whole of the county north of the Irvine. It is much the most populous, and a larger proportion of its surface is cultivatable than of the other two, and it is the most fertile; its whole area is about 185,000 acres, of which it is estimated about fifty-seven per cent is under cultivation. The land rises from the sea-border by easy declivities, and terminates in the pastoral and moorland county of the eastern boundaries. Kyle occupies the central portion of the county; its boundaries are the waters of the Irvine and the Doon. Its area is about 270,000 acres, of which about forty per cent are under cultivation. It is less fertile than the Vale of Cunningham, and more hilly. Carrick, or the rugged, extends from the Doon Water to the southern boundary. This division is generally hilly, with a few fertile and productive valleys. Of its estimated area of 280,000 acres, but thirty-four per cent are

---

[2] Archibald Sturrock, Pr. Essays High. Soc. 4th ser. i, 24.
[3] Archibald Sturrock, *op. cit.* p. 21.

under cultivation. Unlike the other districts, Carrick is as yet almost exclusively agricultural and pastoral.[4]

The climate of Ayrshire is said to be the most humid in Scotland. The winds blow from the west and southwest for more than two thirds part of the year, and the rains from these quarters are frequent, often copious, and sometimes of long duration.[5] The rain does not usually fall in heavy, casual plumps, but comes down in more continuous succession of steady, moderate showers, or thick, drizzling smirrs.[6] This is well-shown by a series of statistics of the rain-fall in Kilmarnock, from March to October, during the years 1864 and 1865. The average weekly rain-fall was .63 inch, and in but five weeks of the sixty was no rain-fall recorded. During the fifteen years from 1850 to 1865, out of the 214 days from March to October, on the average, 109 were recorded as wet.[7] This constant moisture is favorable to the grasses, and is an encouragement to dairy industries.

The temperature is remarkably equable, the colds of winter being mitigated by the passage of the prevalent winds over the adjacent seas, and the extreme heats of summer, in like manner moderated through the influence of the water. During the season of growth, the mean maximum and minimum temperature of any week seldom varies more than 25°, and rarely does the mean maximum attain 65°. The

---

[4] Archibald Sturrock, *op. cit.* p. 21.
[5] Aiton's Survey of Ayrshire, p. 18.
[6] Sturrock, *op. cit.* p. 27.
[7] Thomson's Pr. Essays High. Soc. 4th ser. 46, 347.

mean temperature given for the neighboring city of Glasgow[8] for the year is 47°, and this may be assumed for the temperature of Ayrshire.

The soil is mostly clay in the arable portions. Sturrock estimates more than half of the arable lands to be clays and heavy loams. The clays on the higher ridges are thinner and nearer the till, of a brownish-red color generally, and totally unworkable for green crops under their climate. That kind of clay soil hardens into a brick-like substance during the occasional summer droughts. As for level " carse clay land," there is none in Ayrshire. The light land is comprised mostly in a strip extending along the coast, in an almost unbroken line from the northern boundary to the Girvan River, from one to three miles wide, and perhaps fifty miles long, close to the coast, uncultivated for the most part, but improving in quality as it extends inland. Considerable extent of deep, light loam occurs through Kyle and Cunningham, on the banks of rivers, and more, even of a finer quality, in some of the minor vales of Carrick. Large areas of peat and moor land exist; and although at times some effort has been made for its improvement, but little has been done for the past thirty years.[9]

The principal crops of the county are grass, oats, and wheat. Of these, grass occupied about 57 per cent of the rotation in 1857. About 24 per cent of the average was in oats, and about 6 per cent in

---

[8] Blodgett's Climatology, p. 54.
[9] Sturrock, *op. cit.* pp. 25 and 26.

wheat. If we class the products under white crops, so called, such as wheat, barley, oats, etc., and green crops, which include turnips, potatoes, beets, etc., we have about 31 per cent of the average under rotation for the first, and 11 per cent for the latter.[10]

The dairy is the principal interest, although grazing is carried on to quite a large extent. In 1866 there were 5.7 cows for every hundred acres of area, and 4.6 of other cattle,— a total of 10.3 per hundred acres for neat stock. About 35 sheep and 2 pigs are kept for each hundred acres of area, or a total of 176.9 head of live stock (not including horses) per hundred acres of area of the county.[11]

It is thus seen that a general description of the county is a semicircle of arable land, surrounded by hills suitable for pasturage, there being a natural distinction between the tillage and pasture land. Owing to this basin-like character, from certain elevations more land under culture can be seen at one time than in any other county in Scotland.

The population of the county in 1861 was 198,971.[10] It contains valuable mines of coal and iron, which give employment to large numbers, and as a manufacturing district it stands next in importance to the contiguous counties of Lanark and Renfrew. It is accordingly well supplied with home markets, and its nearness to the city of Glasgow has a favorable influence on its prosperity.

---

[10] Enc. Brit. xix, 797.
[11] Jour. R. A. S. of Eng. 1866, p. 426.

The first definite reports we have of the agriculture of Ayrshire embraces the period comprehended between the years 1750 and 1760. Colonel Fullarton, writing in 1793, states that at this time there was hardly a practicable road in the county. The farm-houses were mere hovels moated with clay. The few ditches which existed were ill constructed, and the hedges worse preserved. The land was overrun with weeds and rushes, and gathered into such high, broad, and serpentine ridges, interrupted with baulks, that a man was required, armed with a pole hooked to the beam of the plough, to regulate the width of the furrow, a device rendered necessary by the extraordinary height of the ridges, some of them being nearly at an angle of 30°. The soil was collected on the top of the ridge, and the furrow drowned in water. There were no fallows, nor green crops nor sown grass. The ground was scourged with oats succeeding crops of oats, as long as the harvest would pay for the seed and labor, and afford a small surplus of oatmeal for the family; then after a period of sterility, or overrun with thistles, it was called upon for another scanty crop.

The farms were of small size, and occupied by mixed tenants, and were divided into what were called the croft or infield, and outfield land. The croft, which was a chosen piece of land near the house, received all the dung, which was of small avail, and which the farmers dragged to the field on cars or sledges or tumbler-wheels, which turned with the axle-tree, and were hardly able to draw five hundred weight. After several crops of oats, a crop of bigg, or four-rowed barley, was taken. Then remaining in lay a year, the land was again broken up to undergo the same wretched rotation. The outfield was kept in a state of absolute reprobation. It was cropped with oats and grass, without dung or other manure.

As there were few or no enclosures, the cattle were either tethered or herded during the summer months, and from the end of harvest, till the ensuing seed-time, were suffered to poach the fields. Starved during the winter, they were scarcely able to rise without aid in the spring, and perpetually harassed during summer, were never in a fit condition for market.

The state of the markets was so low, and so little public credit established, that no tenant could command money to stock his farm, and few landlords could raise the means for improving their estates.

The consequences of this mismanagement were deplorable. The people, having hardly any substitute for oatmeal, were entirely at the mercy of the season. The price of meal fluctuated, and in unfavorable seasons dearth or famine unavoidably ensued. About the year 1700 there were a succession of bad seasons, which reduced the county of Ayr to the lowest gradation of want, and hundreds of families had to fly for subsistence to the north of Ireland. In these seasons of misery, the poor people not unfrequently have been obliged to subsist by bleeding their cattle, and mixing the blood so procured with what oatmeal they could obtain.[12]

At this period, the farmers were altogether ignorant of the fundamental principles of agriculture, and were so much preoccupied with mysterious and abstruse points of systematic divinity that they sought for no other knowledge; and the time which should have been spent on the farm, was occupied in the labors of reform, in demolishing churches, and hunting down the popish clergy, who were the best farmers then in Scotland. A good crop they imputed to the favor, and a bad one to the frowns of Heaven, and, knowing nothing of the principles of vegetation, sought their agricultural returns by greater sanctity and longer prayers, in the place of that labor which springs from understanding.[13] Innovations were resisted. The introduction of a winnowing machine was noticed from the pulpit, and prejudice fostered against it, even to the extent of calling it the "De'il's wind." Accidents happening to those seeking agricultural reform were considered special providences, expressing the disapprobation of Deity.

Aiton, in speaking of the enclosures of the county, remarks that there were no dykes in Ayrshire till about the year 1750, and very few till after 1760. Nine tenths of the fences have been formed since 1766.[14]

Mr. Robertson, one of the ministers of Kilmar-

---

[12] Quoted in Aiton's Survey of Ayrshire, p. 69. See also a similar account, Farmers' Mag. vol. 15, p. 173.
[13] Aiton. *op. cit.* p. 74. Read, also, chap v, vol. 2, of Buckle's History of Civilization in England.
[14] *Op. cit.* p. 221. See, also, Farmers' Mag. vol. 15, 173.

nock, says that about 1760 no enclosures were to be seen, except perhaps one or two around a gentleman's seat, in all the wide-extended and beautiful plain of Cunningham. Here at the end of harvest, when the crop was carried into the barn-yard from the fields, the whole county had the appearance of a wild and dreary common, and nothing was to be seen but here and there a poor barn and homely hut, where the farmer and his family were lodged. The cattle roamed at pleasure and poached all the arable ground, now saturated with the winter rains, so that it was spoiled for the crop of the following year.[15]

Yet there must have been some exceptions to this account, although it probably describes the general state of the county. The parish of Dunlop appears to have been distinguished agriculturally as early as 1700,[16] and in 1740 a Mr. Boyd purchased a cow at the then unprecedented price of £2 2s.[17] The fact of people coming from a considerable distance to obtain a sight of such a famous animal would indicate that the seeds of progress were dormant, rather than dead, in the community, and that occasional improvements or efforts towards change must have been taking place.

In 1804 we find all the wretchedness changed. "Were a person now to stand," says a writer in the "Farmer's Magazine" of that year, "upon an eminence, and survey the beautiful plains of Kyle and Cunningham, with a considerable part of Carrick, he

---

[15] Farmer's Magazine, 1804, p. 73. [16] Forsyth's Beauties of Scotland, ii, 439.
[17] Aiton, *op. cit.* p. 172.

would see the hedges, belts, and clumps of trees already grown to considerable height, fields brought into regularity and order, and spirit and activity everywhere displayed upon something like systematic principles."

The change is further seen by the rent and value of lands at these different periods. The rent of the whole parish of Ardrossan was about £603 in 1749, £3,433 in 1795, and £6,098 in 1808.[18] In Grougar parish, Aiton gives the valuation of one piece of 70 acres at £170 in 1742, and £7,000 in 1811. The whole arable lands of the parish of Kilmarnock were placed in 1763 at 2½ to 3 shillings per acre; their rental in 1811 was twenty times that sum.[19] Yet during this time the price of wheat, taken in average periods of ten years, had changed but very little; the price of bear or barley had advanced greatly, while there was a steady advance in the price of oats and oatmeal.[20] But little wheat could have entered into the consumption of the people, for until the year 1785 but little was seen beyond the limits of a nobleman's farm.[21] The increase in the price of the staple products of oats and barley could not have justified the increased rents, were it not for the increased production.

It may be well to inquire into the causes for this change. The atrocious religious persecutions had left the country at the close of the seventeenth century in a bad state, and had imbued the people of the earlier

---

[18] Aiton, *op. cit.* p. 168.
[19] Aiton, *op. cit.* p. 169.
[20] Aiton, *op. cit.* p. 171.
[21] Gazetteer of Scotland, 1, 90.

portions of the eighteenth century with a religious fanaticism, which hindered progress, and bordered on criminality in its interference with the development of the country.[22] With the revolution of 1688, a new era commenced in the legislation on corn, and soon after in the practice of the cultivator in Britain;[23] and the greater attention paid to improvement, as following the tendency of the times, was not without its effect in Scotland. In 1723 a society was formed for the Improvement of Agriculture, of which the Earl of Stair was a most active member; but there is reason to believe that the influence of the example of its numerous members did not extend to the common tenantry.[24] It is worthy of remark that farmers are at all times tardy and reluctant in following the example of those possessing wealth; whereas, when a person who depends upon the success of his industry for his subsistence, prospers in his pursuits, his example is quickly followed by others in his neighborhood.

The County of Ayrshire contained within itself, however, the elements of reform; and Alexander, Earl of Eglinton, commenced the improvement of his large estate about 1730. He spurred the industry of his tenants by personal appeals, opened quarries, laid off roads, plantations, and ditches, and introduced an eminent farmer from another district. John, Earl of Loudon, also began extensive improvements about this time, and raised field turnips, cabbages, and car-

---

[22] See Buckle's Hist. of Civilization in Eng., vol. 2, chap. 5.
[23] Enc. Brit. ii, 254.   [24] Enc. Brit. ii, 262.

rots as early as 1756.[25] But the most fruitful stimulus for improvement were the Acts of Parliament between the years 1750 and 1760, for collecting tolls and making roads.[26] It is not easy to estimate the benefit which agriculture has derived from good roads, and the want of communication was one of the causes of the slow progress of the art in former times.

About this time the Earl of Eglinton established a Farmers' Society, and presided over it himself for a number of years.[27] The gradual advance in price and produce, the consequence of increase of population and manufactures, giving a powerful impulse to rural industry, rendered possible the changes in the system of leases and the restrictions on cultivation and rotation. The Fairlie rotation, introduced by the Earl of Eglinton, was pursued by William Fairlie, after this nobleman's death, not only upon the Earl's extensive domain, but also on a considerable property of his own.[28] Every farm as it came out of lease was enclosed, and divided by sufficient fences into three or more parts, and was allowed to remain in grass till it recovered from the exhausted course of evil management already stated. The land was limed, convenient houses and offices were builded, and a lease granted, usually for eighteen years, under covenant not to plough more than one third of the farm in any one year, nor to plough the same land more than three successive years. With the third crop, the land

---

[25] Aiton, *op. cit.* p. 80.
[26] Enc. Brit. ii, 262.
[27] Aiton, *op. cit.* p. 678.
[28] Farmers' Mag. 1804, p. 783. Aiton, *op. cit.* p. 81.

was laid down to grass. The fodder was stipulated to be consumed upon the farm, and all the dung to be spread upon it.[29] Other proprietors followed in these courses, and the increased rents which such measures demanded had the tendency to drive out the shiftless farmers and replace them by men of energy and force. It is obvious, as Aiton observes, that many of those who pay the highest rent realize the largest profits. This proceeds from the increased industry to which they are roused, by knowing that they have a higher rent to raise. And many of those whose rents are extremely moderate, as well as some of the small proprietors who pay no rent whatever, have by their indolence been reduced to poverty. Some proprietors within his knowledge, having gone through bankruptcy and sold their land, occupying it as tenants at high rents, have gained by industry as tenants double the sum which they had obtained as the price of their own property.

In 1786 the Kilmarnock farmers established their Society, and a few years after others were formed at Maybole, Galston, Newmills, and other places.[30] From 1784 to 1795 the improvements advanced with steady steps. From 1795 to 1814 the prices of produce steadily increased under the influence of the Napoleonic wars. The date of 1784 is that of the origin of the Highland Society, whose prominent objects were then stated to be to facilitate communication and advance agriculture; and their list for the

---

[29] Aiton, *op. cit.* p. 85.
[30] Aiton, *op. cit.* p. 680.

year 1789 offers premiums for essays on the management of cattle-farms and the dairy, breeding stock, etc., for the execution of improvements, and the raising of crops."[31]

It was, therefore, the extension of the activity of thought following the political measures of this time, and introduced into the common life of Ayrshire, which rendered possible these sudden changes. It is seldom that human agency has effected so much in the environment of a country, as took place with such remarkable activity in Scotland : which changed a waste into a garden; which furnished such contrasts between what might have been seen by an individual in the course of an ordinary life-time. Although our accounts of the cattle of this district are few, it cannot be doubted but that the appearance and properties of the cattle and the dairy shared in these contrasts, and the changes which were possible in the tenancy of the land, were possible in a breed; and the changes which actually took place in the one, must have produced a change in the other.

---

[31] Prize Essays, High. Soc. vol. 1.

AYRSHIRE CATTLE: DOCUMENTARY HISTORY OF THEIR ORIGIN.

Breeds of cattle attain their excellences and their prominence by degrees, and their early history is difficult to be traced, as each addition to their usefulness has either been unrecognized, or has seemed at the time too insignificant to record.

The history of the Ayrshire breed of cattle is shrouded in the past. But few efforts have been made to lift the veil, and the scanty records that we have, seem little better than personal surmise or unfounded assertion. Everything beyond what well-attested accounts reach is obscure, and the more so here, as few writers have considered the history of agriculture in its details, or the occurrence of a well-defined or growing breed, worthy the pen of the historian.

Of one fact we are certain. About the close of the last and beginning of the present century, our attention is called to a breed of dairy cows established in the County of Ayrshire, and already having a local celebrity for the quantity and profitableness of their yield of milk. Their origin was probably influenced by the general revival of agriculture which took place in the eighteenth century.[1] It is a peculiarity of the human mind to desire to fix a definite origin for a race or a man in whom there is a strong interest. The early Greeks recognized the obscurity of beginnings, and accordingly derived the origin of their heroes from a divine progenitor. In a like manner writers on cattle have attempted to derive the origin of their breeds, from imported animals or obscure crosses. They have attempted to use the divinity of a recognized breed in support of the breed, which they fear will seem to their readers comparatively recent. The literature of the Ayrshire breed abounds with this error.

Aiton, our principal and almost only authority on the origin of this breed, understands that the Earl of Marchmont, about 1750, purchased from the Bishop of Durham and carried to his seat in Berwickshire, several cows and a bull of the Teeswater or other English breed, of a brown and white color, and that some of this breed were carried to Sombeg, in Kyle, and crossed with many cows about Cessnock and Sundrum.[2] John Dunlop, of Dunlop, is also said to have introduced cows of a large size from a distance, probably of the Dutch, Teeswater, or Lincoln breed.[2] In a later writing, Aiton, laboring under a seeming necessity of giving a more definite origin to the breed, writes that about 1770, or a little earlier, bulls and cows of the Teeswater or Short-horned breed are "said to have been introduced into Ayrshire by several

---

[1] Burton's Scotland, ii, 393.
[2] Survey of Ayrshire, p. 424.

proprietors, and it is from them and their crosses with the native stock, that the present dairy breed has been formed."[3] When writing in 1811, however, he says that the Ayrshire dairy breed is "in a great measure the native indigenous breed of the County of Ayr, improved in their size, shapes, and qualities, chiefly by judicious selection, cross coupling, feeding, and treatment, for a long series of time, and with much judgment and attention;"[4] and this appears from the context a more correct expression of his judgment, and the fact, than the other.

When we pass to general statements of their origin, we find the author of the "Complete Grazier" asserting them a cross of the Alderney cows with Fifeshire bulls, under the name of Dunlop breed.[5] Ro. Forsyth, however, writing in 1805, speaks of the Dunlop breed as having been established in the parish of that name for more than a century.[6] Quayle, who wrote the "Agricultural Survey of Jersey," states that the Ayrshire breed was a cross between the Short-horn and the Alderney,[7] and Col. LeCouteur, of the Island of Jersey, writes[7] that Field-Marshal Conway, the Governor of Jersey, and Lieutenant-General Andrew Gordon, who succeeded him, nearly half a century back, both sent some of the best cattle to England and Scotland. Ro. Forsyth, again, that elegant and apparently trustworthy writer, says[8] that the Earl of Fife, and General Grant, of Banffshire, have spared

---

[3] Sinclair's Scotland, 1814, iii, 43.
[4] Survey of Ayrshire, p. 422.
[5] Complete Grazier, 3d ed. p. vii.
[6] Beauties of Scotland, ii, 439.
[7] Quoted in Jour. R A. S. of Eng. 1844, p. 47.
[8] Beauties of Scotland, iv, 456.

no expense in introducing from time to time the most valuable breed of bulls and cows from England and Germany. As the Duke of Gordon had his family seat in this shire, and as the dates of the two statements agree, it is possible that they refer to the same event. John Orr, Esq., of Barrowfield, brought from Glasgow, or some part of the East County, to Grougar, about 1769, several very fine cows,[9] which fact would seem to show an occasional movement of improved stock from distant districts.

The cattle of this district, at the time we have our first accounts, were black and white. Indeed, so common was this color that Cully remarks, that in all the accounts of cattle which he had seen in deeds or statutes, they are called black cattle. Black or brown with white faces, and white streaks along their backs, were the prevailing colors in Ayrshire in the earlier portions of the eighteenth century.[10] Aiton describes them previous to 1750, as being generally black, with some white on their face, belly, neck, back, or tail, and in 1811 as mostly of a dark color, or black, with the exception of the improved dairy breed.[11] Again he speaks of them, from his own recollection, as black, with white on the face, the back, and the flanks, and few of the cows yielding more than from one and a half to two gallons of milk in the day at the height of the season.[12] Still later in his writings he states that about 1770, they were of small size, with high-standing, crooked horns,

---

[9] Survey of Ayrshire, p. 424.
[10] Beauties of Scotland, ii, 439.
[11] Survey of Ayrshire, p. 425.
[12] Quoted in Low's Animals, p. 342.

narrow on the back, and flat on the ribs, and mostly of a black color, with white spots on their faces and other parts.[13] These descriptions show an affinity with the Highland breed.

The first record of improved cattle that we are able to find goes back to about the year 1700. Ro. Forsyth, in 1805, states that for the purpose of preparing cheese in Cunningham, a breed of cattle for more than a century had been established, remarkable for the quantity and quality of their milk in proportion to their size. These had long been denominated the Dunlop breed, either from the lands of the ancient family of that name, or from the name of the parish where the breed was first brought to perfection. Our only other reference to this breed by name is in the "Complete Grazier," an anonymous work, published about 1800, where it is stated that the Dunlop breed, is a cross from Alderney cows with Fifeshire bulls, and are described as small in size, of a pied or sandy red color, and with small horns awkwardly set on.[14]

In 1778 and 1780 the color of red and white became fashionable;[15] and between 1785 and 1805 the brown and white mottled cattle became so generally preferred, as to bring a larger price than others of equal size and shape, if differently marked;[16] and Aiton speaks of the red and white being common in 1810.[15]

Color is affected much by varied conditions; and

---

[13] Sinclair's Scotland, 1814, iii, 43.
[14] Complete Grazier, 3d ed. p. 7.
[15] Survey of Ayrshire, p. 424.
[16] Beauties of Scotland, ii, 439.

oftentimes a change in environment, although almost inexpressibly small, as illustrated on the cattle-farms of the Pampas,[17] will cause the self color of a wild or semi-wild breed to break; and when the varied conditions accompanying agricultural improvement reached the cattle of Ayrshire, we find by our record a greater change in the colors than had existed under the less varied circumstances of agricultural stagnation.

As the spirit of travel and improvement reached the upper class of inhabitants, we find the merits of foreign breeds recognized, and an introduction of other breeds, to a sufficient extent, at least, to vary the color-marks of the cattle; and those colors which became fashionable, and thus sought after with greater avidity, would naturally become the most general. We thus find at the present day the red and white preponderating over the other colors, and the blacks and whites far less common than in the past.

Ro. Forsyth, not realizing the quick changes produced by the directing of general attention to certain points, as profitable or fashionable, remarks upon the rapidity of the diffusion of the improved breed, as a singular circumstance in the history of breeding, and speaks of the mottled breed as of different origin from the common stock.[18] He describes this variety in 1805 as being short in the leg, with fine-shaped head and neck, and small and tapering horns, their body deep but not so long, nor so full and ample in

---

[17] Azara, Quad. of Paraguay.
[18] Beauties of Scotland, ii, 439.

the carcass and hind-quarters, as some other kinds.[19] This description has a bearing on the origin of this breed, as it shows that at this date no change had been produced which could not be accounted for by selection and treatment.

---

[19] Beauties of Scotland, ii, 439.

ORIGIN OF AYRSHIRE CATTLE.

We have seen that cattle abounded in Scotland before the historical epoch; and throughout the earlier centuries, the pasturage of herds and the manufacture of cheese are recorded fragmentarily and concisely, in the charters and excerpts of the monks and earlier historians. We have it stated by a competent writer that a breed existed in Dunlop, a parish of Cunningham, as early as the year 1700, which was noted for quantity of milk in proportion to size; and the same writer gives it as a veritable fact that a certain Barbara Gilmour, fleeing the county to escape the barbarities accompanying the religious persecutions, under the last princes of the house of Stuart, introduced upon her return from Ireland the manufacture of cheese, which since that period has been the great business of that neighborhood. He proceeds, "Sensible that their situation was more favorable for this than for any other purpose, the people bestowed upon it the greatest care and turned it to the best advantage."[1] In this sentence we have the key to the origin of the improved breed.

In the region which included this county we have records of the earliest attempts at civilization; and in

---

[1] Ro. Forsyth, Beauties of Scotland, ii, 441.

DIFFERENTIATION. 145

the differentiation brought about by its consequences, we find the cattle lacking an uniformity of color, yet in many respects resembling the breed which formerly inhabited the wilds, and which now, degenerate, inhabits the parks of certain nobles. These cattle differ from the wild herds in color, but this, it is shown, is hardly an important character, as the wild cattle display a strong tendency to vary among themselves. Moreover, the improved breed occasionally sport into white with red ears, resembling thus the forest breed.[2] The breeds of the county have not the heavy mane, which history and tradition have ascribed to the forest animal; yet this animal has lost it wholly or in part.

As ideas of agricultural improvement reached these regions, there is evidence of increased interest being taken in the breeds; the more obvious feature of color is taken in hand, and brown and white colors are preferred, and the result is a remarkably rapid diffusion of these colors throughout the district during the years intervening between 1785 and 1805, the era of the Agricultural Society, and the certain identification of the improved breed. The more spirited of the agricultural improvers, attracted by the fame of foreign breeds, introduce now and then

---

[2] U. S. Pat. Off. Report, Ag. 1851, p. 91, note.

In October, 1872, a white heifer-calf, with ears tipped with red, was dropped, from Ayrshire parents, at the Massachusetts Agricultural College. In October, 1874, we saw in the farm-yard of Mr. Tilton, at Martha's Vineyard, two cows, perfectly white, save the inside of the ears, which were brown-red two thirds down from the tip. These animals were the result of a cross, an (grade?) Alderney bull and grade Ayrshire cow, — a good illustration of reversion brought about through crossing. In appearance these animals resembled Landseer's picture of the White Forest Breed, — probably the only two white animals on the island.

such animals, and rear these, or cross with the native stock, and during this whole time a process of selection for uses is going on by all alike, — the cow giving the most milk being retained, while the poorer milker finds her place in the shambles. The progeny of the largest milking animal is reared, in preference to others whose ancestors are not so well, or unfavorably, known for this quality. The fashion and the natural eagerness to secure those colors which are attractive, also come into play; while the improved system of farming, the enclosing of lands, the winter protection, and other adjuncts of improved agriculture, aided in bringing the breed to a larger size and greater excellence.

It is possible that the Guernsey breed may have transmitted some of her quality to the present Ayrshire, as is suggested by the sandy red and pied Dunlop; but if so, it is scarcely ever shown at present in color of skin or hair. Similarity of function can produce a certain similarity in form; and whatever resemblance may exist between the Alderney and the Ayrshire can well be referred to this law. We find a correlation between the external parts of a cow and her physiological functions; and two separate peoples, seeking in a breed dairy qualities, would naturally and unavoidably obtain certain shapes in common, from whatever breed they may have originally started. It is in the point where differentiation occurs that we would look for divergence, and we see it in the udder: the one breed designed for butter alone, the shape of the udder is neglected in the breeding, and we see the pointed, egg-shaped, and goat udder almost universal; the other breed designed for milk, and the udder is admired for its capacity; and we therefore find it broad, more level on its sole, and extending far forward and back.

It is possible, even probable, that Short-horn crosses may have occurred; for it would be strange that a breed so well and favorably known could exist so near the region of Ayrshire, without attracting the attention of wealthy gentlemen, who were desirous and eager to advance the capabilities of their heritage.

The Holderness, said to have been introduced into the north of England and south of Scotland,[3] also may have been used in modifying the breed; and it is highly probable that the indistinct black spots which occasionally show through the white hair of the Ayrshires, may be accounted for under the laws of reversion. Where so little is known with certainty of the origin of a breed, and where recorded instances of the presence of other breeds are given, the probabilities of a mixture become almost certainties. The presence of cattle from the Irish coast, in the adjacent island of Arran, and the introduction of these same cattle into Galloway,[4] would seem to afford a reasonable presumption of crosses having occurred with these animals in the region of Ayrshire. It is possible that the orange rim to the eye, occasionally met with among the Ayrshires, is derived from a distant Kerry ancestry.

---

[3] Low's Animals, p. 380.   [4] Youatt on Cattle, p. 75.

The Ayrshire breed is undoubtedly the descendants of the original wild breed, modified by civilization, and more particularly by selection; and the selection has certainly been aided by the variations produced by crossing with other and distinct breeds. Improvement, as thus begun, was probably at first local, then gradually extended, until the enclosure of the fields, and the demand for certain produce, increased the number of the areas of local betterment. These agencies, acting for a long time, but more particularly within the period comprised in the last of the eighteenth and first of the nineteenth centuries, resulted in an animal of varied markings, but distinct quality; and in 1810 we can claim the existence of an improved race, of remarkable dairy capacities, so well bred in as to be permanent for the breed.

The origin of the Ayrshire breed is, in a word, adaptation. The united efforts of the spirit of improvement, and the influence of locality, acting on such materials as were at hand, and guided by an unconscious selection, acted on by a general intelligence, produced an animal which is a determinate product, of an age characterized by a special activity in promoting progress. The same agencies which evolved the steam-engine into usefulness had a part in evolving the improved Ayrshire cow.

PROGRESS OF IMPROVEMENT SINCE 1805.

The Ayrshire cow of 1805, although possessing some fineness of shape, and credited with a not uncommon yield of from 24 to 34 quarts of milk daily, and exceptionally as giving as much as 40 quarts, yet appears to have been deficient in width and depth of carcass behind, and no mention is made of the shape either possessed by, or desired in, the udder. This breed, however, was very generally diffused over Cunningham,[1] and very soon found its way into other counties of Scotland.[2]

In 1811, as we judge from the description and figure given by Aiton,[2] the shape of the carcass had somewhat improved, and there seems to have been gained a lightness forward. At the same time our attention is called to the shape of the udder, which is described as broad and square, stretching forward, neither low-hung nor loose. The same stress is laid on the perfection of udder in the description given by William Harley, in 1829,[3] and he had cows which not unfrequently gave from 25 to 30 quarts a day, and once even attained 40 quarts.

The great breadth and depth of the loins appears

---

[1] Ro. Forsyth, Beauties of Scotland, ii, 439.
[2] Survey of Ayrshire, 426.   [3] Harleian Dairy System p. 106.

to have been gained in 1845,[4] and now, also, we first find mention of the flatness of thigh, at the inner side technically called the twist. At this time, the drooping of the haunch towards the rump was common. This breed had now become the prevailing stock in Renfrew, Dumbarton, Stirling, and Lanark, and had been carried to many other more distant localities.

In 1853[5] we have for the first time a recognized standard for the breeder, the Scale of Points of the Ayrshire Agricultural Association. Particular stress is placed on the wedge-form body, and the development of the rear half of the body where the concentration of function takes place. The shapes of the milk-vessel and its appendages receive greater attention, and there is demanded an increased fineness of points.

In 1866 Archibald Sturrock, in a prize essay on Ayrshire County, writes that "a capacious and well-set udder is certainly the chief point of excellence."[6]

In 1868 the chief point of merit of Ayrshire cattle is said to be "a capacious and well-set udder, and these are the principal objects aimed at, although a straight back, with a sweet head and branching horns, are received with favor in a show-yard."[7]

In 1871 a writer in the "Farmer's Magazine," in describing the Ayrshire cow, proceeds: "The udder well set on. For a prize-taker this point must be faultless, as no beauty of form or regularity of other

---

[4] Low's Animals, p. 343.
[5] Pr. Essays H. Soc. 1866–7, p. 106.
[6] Pr. Essays H. Soc. 1866–7, p. 77.
[7] H. N. Fraser, Pr. Essays H. Soc. 1868–9, p. 331.

points will make up for deficiency in the form or size of the milk-vessel. If this is in perfection, other and minor points may be overlooked."[8]

The most noteworthy fact in the above series is the stress laid upon the form of the udder, and this has been caused by the educating influence of the many farmers' clubs, with which the district has been sprinkled. This influence was early manifested, and competition must have had a great influence, in changing the form of this useful portion of our animal, into a vessel not alone for use, but for beauty.

In 1836, a large premium was offered for the competition of this breed, by the Highland and Agricultural Society, which long before had offered encouragement to breeding stock; and the local societies, some of long antiquity, had so increased, that in 1866 each parish had its local society, in addition to "estate clubs,"[9] while the county society supplemented the efforts of the smaller unions by embracing the whole area and giving more weighty encouragement.

The effect of this interest in the breed, was to incite the farmers to stronger efforts towards improvement. The leading type of the breed at one time, is said by Sandford Howard to be of the Kyloe or Highland cross, and he vouches for the facts obtained by himself, substantially as follows: "Theophilus Parton, of Swinly Farm, near Dalry, Ayrshire, about forty-five years ago [1818] took great pains to establish a

---

[8] Quoted in Nat. Live Stock Journal of Chicago, Feb. 1871, p. 183.
[9] Pr. Essays H. Soc. 1865-7, p. 75.

herd of what were deemed the best Ayrshire cattle, into which he infused a strain of the West Highland blood, the particular degree of which is not publicly or generally known. The Swinly stock differs from the older Ayrshire in having a shorter head, with more breadth across the eyes, more upright and spreading horns, more hair, and that of a more mossy character, and generally better constitutions. They are also somewhat smaller-boned than the old stock, though from their superior symmetry and greater tendency to fatten they are fully equal to the former in weight of carcass when slaughtered." [10]

In 1847 the St. Quivox Club attempted to introduce the Short-horn breed more generally among breeders, but it failed to produce any effect, as we are told by Sturrock, as now "Short-horn crosses are more difficult to procure than formerly." [11] Professor Norton, of Yale College, speaks of seeing, during a visit to Scotland in 1848, Short-horn bulls on every large farm, but leaves the inference that the crosses were designed for beef. [12]

Mr. Wilson, in writing of the agriculture of Lanarkshire, states that this cross, although it diminishes the milk, yet adds increased value for the shambles. [13] These statements, taken together, seem to indicate that Short-horn crosses were used only when grazing was united with dairy farming.

In conversing with the breeding farmers of Ayrshire in 1869, we were unable to find any Short-horn

---

[10] U. S. Dept. Ag. Report, 1863, p. 195.
[11] Pr. Essays H. Soc. 1866-7, p. 37.
[12] Farmers' Lib iii, 306.
[13] Prize Essays H. Soc. 1836-7, p. 365.

crosses, although some Short-horn bulls were found on farms uniting the business of the grazier with that of the breeder. The black color was referred by some to the influence of the Highland race. It would seem as if parentage would occasionally crop out in the colors; and although a red might now and then suggest a Devon, or a brindle or black the Kyloe, and rarely a pale red the Alderney, yet we saw not a single roan which would indicate the Durham. Of the Jersey bull or cow we saw not a trace, and our inquiries provoked the curiosity which indicated an unfamiliarity, even, with the appearance of that breed.

The Highland cross appears to have been frequently used, especially by those who desired handsome bulls. A Mr. Horne, in remarks before an agricultural club in 1867, states, from his own observation, that a famous prize-taking bull, Geordie, was popularly accounted to have an eighth of West Highland blood.[14] This cross gives a style to the carriage of an animal, and increases the tendency towards laying on flesh or fat.

Perhaps the history of the progress of the Ayrshire breed since 1810 can be best summarized by an extract from a letter written by a prominent and careful breeder of more than ordinary intelligence, Robt. Wilson, of Forehouse, Kilbarchan:—

"Modern Ayrshire cattle have been brought to their present condition by care and attention on the part of breeders,— each selecting according to his

---

[14] Gard. Chron. and Ag. Gaz., July 27, 1867.

fancy, and crossing accordingly. There is no doubt but the majority of Ayrshire cattle have been crossed, as distinct points of Highland, Short-horn, Devon, Hereford, etc., are easily discernible, not only in color but also in style. *In dairy districts*, however, the pure breed is *invariably* attempted to be kept, and crossing, therefore, is more the exception than the rule." He adds that the breed has not improved in some respects within his remembrance, but that "so far as the fine points are concerned, probably the number of fine-bred cattle is greater than ever before."

III.

LOCAL.

IN preparing the list of importations and importers, we had fondly hoped to obtain our information at first hand, from the parties at interest, — the importers themselves. We therefore circulated very freely a printed request for this and other information. We received some really valuable replies, but the largest portion of our broadcast circular sowing fell on barren ground.

Whatever errors, therefore, are found in the list, may be charged to the indifference of those who, at first thought, would be supposed to be the most concerned in its accuracy. By presenting our authorities in every case, except where we have had private information, we can avoid the charge of carelessness, and say we hope to have attained very considerable accuracy.

In preparing a list of animals that have taken prizes at Scotch fairs, our intention is to show certain animals which may be considered as thoroughbred. Although many animals have been imported which are as truly thoroughbred as some which have received premiums, yet the mere fact of importation cannot be a guarantee of authenticity. Some true Ayrshires have been imported from localities far removed from their own county, and some inferior or uncertain animals have found their way here under the impulse of speculation. Prize-taking in Scotland is one guarantee of authenticity.

We shall have to ask breeders for charity towards the imperfections of the list. It results through their own negligence.

The matter of pedigree must be considered one of the greatest importance. It is the Alpha and Omega of breeding. It must be sought for continually, retained pertinaciously, and intensified yearly, in order to achieve the greatest success. We therefore present a few thoughts under this heading.

IMPORTERS AND IMPORTATIONS.

As early as 1822, or thereabout, we find record of the introduction to America of this useful breed. In that year a bull and a cow are said to have been brought from Great Britain to New York by Mr. Henry W. Hills, and sent to the farm of Mr. Hezekiah Hills, at Windsor, Conn. The cow was afterwards sold to Joseph Morgan, of Hartford, and the bull to Elihu Wolcott, of East Windsor Hill. Two of the heifers, called Flora Hills and Fanny Hills, were sold to Mr. Henry Watson, of East Windsor, "which produced several calves from his Short-horn bull, Wye Comet." These calves, half Short-horn and half Ayrshire, were small animals but very fine, and several of them "were recorded as Short-horns, in the American Herd Book."[1]

According to the "Turf, Field, and Farm,"[2] the Ayrshires were first introduced into this country in 1828. In 1831 we find note of a full-blood Ayrshire cow being in the possession of a Dr. White, of Dutchess County, N. Y.; this cow was crossed with a Durham bull about this time, and then bred in,

---

[1] Samuel Bartlett, in "Homestead," quoted in Rept. Conn. Board of Agriculture, 1867, p. 151.
[2] Quoted in Nat. Live Stock Journ., May, 1871, p. 303.

with her descendants, for a dozen or fifteen years at least.[3]

1837 In 1837, their merits at home having become more widely known, we learn of two importations: the one of Mr. J. P. Cushing, of Watertown; the other by the Massachusetts Society for the Promotion of Agriculture.

Mr. Cushing's importation was made in the spring, and consisted of four cows, — Flora, Juno, Venus, and Cora.[4] Three heifers appear to have been imported in their dams,[5 and 6] and perhaps a bull.[5] Some dozen years later Mr. Cushing presented one of his bulls to the Worcester Co. Agricultural Society.[7]

During this year arrived the first importation of the Massachusetts Society for the Promotion of Agriculture, consisting of a bull and three cows, which were all in calf when they arrived. The bull was sent to the western part of the State, and was kept near Pittsfield.[8] One of the cows was placed in the care of Hon. P. C. Brooks, in Medford; another in the care of Hon. Daniel Webster, at Marshfield; and the third of Elias Phinney, of Lexington. This last, 18 years old, was still living in 1847.[9]

In 1845 this Society made its second importation, consisting of a bull, Prince Albert, and four cows, Flora McDonald, Jennie Deans, Milly, and Charlotte. These animals were selected by Mr. Alexander Brick-

---

[3] U. S. Pat. Off. Reports, 1851, p. 91. Note.
[4] Farmers' Lib iii, 304.
[5] Capt. Randall's Ms. Herd Book.
[6] A. H. B., B. 53, 702; C. 661, 732.
[7] Ag. of Mass. 1853, 311.
[8] U S. Dept. Ag. Rept. 1863, p. 197. Cultivator, Feb. 1848, p. 42. Farmers' Lib. iii, 304.

ett, of Lowell, who was sent out for that purpose. They were shipped about the first of September, and landed in Boston about the first of October in good condition, and were placed on the farm of Mr. Phinney, in Lexington.[10]

In 1858, the Society again sent to Scotland, and this time, through Mr. Sanford Howard, selected and imported four bulls and eleven heifers.[11] The bulls appear to have been Tam Sampson, Troon, Albert, and Irvine. Kilmarnock and Young Cardigan were imported in their dams. The cows were Daisy, Gentle, Harriet, Lily, Mavis, Miss Anderson, Miss Morris 1st, Pansy, Rosa, Ruth, and Star. Buttercup was imported in her dam.[12] These animals were from the herds of well-known breeders in Ayrshire. In 1869, while travelling through this county, we stopped at the farm of Mr. John Ritchie, who remembered Mr. Howard well, and stated that Mr. Howard was very particular in his choice, and carried away the best he could buy.

In order to disseminate the blood through the State, the Society at various times presented bulls to the Hampshire and Franklin, Worcester County, Essex, Hampden, Barnstable and Plymouth Agricultural Societies, and in 1849 Jennie Deans was presented to the Middlesex Society.

In 1838, Capt. George Randall, of New Bedford, commenced his series of importations with Maggie,

1838

---

[10] Farmers' Lib. ii, 123. Alb. Cult. 1845, p. 557, etc. 1847, p. 41.
[11] Count. Gent., Feb. 18, 1869, p. 140.
[12] A. H. B., B. 129, 38, 75, 398, C. 32, 40, 83, 129, 138, 155, 167, 187, 195, 199, 424, 643. Ag. of Mass. 1860, pp. 74, 82. Ag. of Mass. 1853, p. 301, etc.

who was landed on the 20th day of July. She unfortunately died the same season. On the second of December arrived the bull Rob Roy, about two and a half years old.

In 1839, on the 27th of July, his third importation arrived, consisting of the yearling bull Roscoe, the four-year-old cow Swinley, and the yearling heifer Daisy. The cow Swinley dropped a heifer calf, Maggie, March 20, 1840, which was sold in 1846 to the Massachusetts Society for Promotion of Agriculture. The heifer Daisy died in September, 1841.

His fourth importation arrived May 26, 1841, consisting of the five-year-old cow Crummie, who dropped a bull calf Wallace, February 2, 1842, and the heifer Daisy. This second Daisy was sent to Capt. Randall as a present by Lawrence Drew, her breeder, on hearing of the loss of the former Daisy.

In 1844, Capt. Randall made his fifth and last importation, consisting of the cow Medal, which arrived September 22, and in the following April gave birth to twins, Sandy and Jeanie.

Capt. Randall's stock was mostly bred by Lawrence Drew, a Scotch breeder well known for his success. They were probably of the best, and the records of these and their descendants were kept with great apparent accuracy and neatness.[13]

1840 In 1840 Capt. Ezra Nye, of Clinton, N. J., seems to have imported a cow, Nan,[14] from the Duke of Portland's estate, Ayrshire.

---

[13] We desire to express here our thanks to Mr. Haskell, of New Bedford, for being allowed to take a copy of Capt. Randall's herd book.

[14] A. H. B., C. 101, 485, 570. Also 1st Rept. N. E. Ag. Soc. p. 57.

In 1845, Capt. Nye appears to have made an additional importation of the bull Duke,[15] and probably the cows Marion, Lily, and Beauty in calf with Scotland.[16] One heifer, Bessie, and four bulls, Antarctic, Leopard, Juniper, and Pacific, are said to have been imported by him, but we find no clew to the date. Our references are certainly misleading unless there were other importations at a later date than these given.[17]

The importation of Capt. George Randall for this year will be found noticed under date of 1838.

In July, 1841, Capt. J. C. Delano brought to New Bedford a cow named Jennie Deans, and about three years old. She was probably purchased partly as a speculation, and in part to supply the ship with milk. She was called pure, but Capt. Randall, into whose possession she afterwards came, had his doubts.

About this year it is said that some Ayrshire cattle were imported by George Longley, of Maitland, Canada.

In 1842, Mr. E. P. Prentice, of Albany, N. Y., imported a cow, Ayr by name. She dropped a heifer calf on the passage, which was called Ayr 2d.[18] The cow is figured in the "Albany Cultivator," of July, 1846.

About this time an importation of animals, selected and forwarded by Allen J. Davie, arrived in Balti-

---

[15] A. H. B., B. 6, C. 70, 101, 570.
[16] A. H. B., C. 70, 162; C. 685; C. 252½, 738; B. 59, 874, 925; C. 14, 113, 131, 192, 278, 1153.
[17] A. H. B., C. 1,382; B. 173, 468; C. 738, 902, 1372, 1443, 1483, 1643; C. 685; C. 162; C. 553.
[18] Alb. Cult. 1845, p. 14; do. July, 1846; do. Feb. 1848, p. 41.

more. These passed into the hands of John Ridgley, and were sent to the celebrated Hampton Estate. This importation was probably kept with but little attention to preserving the breed intact.[19]

For the Randall importation of this year, see the notice under the year 1838.

For the notice of the importation of the Massachusetts Society for Promotion of Agriculture during this year, see under 1837.

For Capt. Nye's importation of this year, see under date of 1840.

Some years previous to 1847 Dr. Hoffman made an importation into Maryland. These passed, some of them at least, into the hands of Mr. McHenry, of Hartford County.[20] Some of this importation appears to be found in the cows Jenny Deans and Mary Queen of Scots.[21]

1846 In June of 1846, R. L. Colt, Esq., of Paterson, N. J., imported a bull and a cow in the ship "Europe." The bull Geordie was a descendant of a famous bull of that name in Scotland, and was himself a prize-taker. He cost £40 in Ayr. The cow Bessy cost £19.[22]

Samuel Ward, Esq., then of North Stockbridge, afterwards of Lenox, Mass., imported a bull and heifer this year.[23] The bull Dandy appears to have gone into the possession of E. P. Prentice.

1848 In 1848, according to a reference in the Ayrshire

---

[19] Farmers' Lib. iii, 1848, p. 286.   [20] Ibid.
[21] A. H. B., C. 1787, 1923.
[22] Farmers' Lib. ii, 385, where the bull is figured; do. iii, 289, for the cow.
[23] Alb. Cult., July, 1851.

Herd Book, R. Jardine imported a cow into St. John, 1848 New Brunswick, and according to other references in 1849. This cow was probably named Gowan.[24]

In 1849, an Ayrshire farmer of the name of R. 1849 Gray emigrated to New Brunswick, bringing with him his cattle. These seem to have consisted of the bull, Jock the Laird, and four cows, Helen, Peerless, Mary, and Jenny Willet.[25] The descendants of these cattle occasionally find their way across the border, and are referred to as being imported from New Brunswick.

Mr. James Brodie, of Rural Hill, New York, has been the active agent in importing for a number of firms, among which occur Hungerford, Brodie & Co., Hungerford & Brodie, Brodie, Campbell & Co., Brodie & Campbell, James Brodie & Son, Brodie, Son & Converse, and Walcott & Campbell. The importations of Mr. G. C. Bradley can also be referred to him.

In 1852 arrived Kilburn and Mary Gray.[26] These 1852 were exhibited at the New York State Fair of that year by Mr. Brodie. Afterwards they came into the possession of Messrs. Walcott & Campbell, New York Mills, N. Y.

In 1853 were imported Ayrshire Lass and White Lilly.[27] The latter was exhibited by Hungerford & Brodie the same year, but some time afterward went to New York Mills.

---

[24] A. H. B., C. 43, 880, 881, 1012.
[25] A. H. B., B. 32, 284, C. 90, 94, 102, 103, 139, 241, 327, 667, 1166, 1228, 1516, 1553, 1554. Also MS. information.
[26] A. H. B., B. 224, C. 592.
[27] A. H. B., C. 236, 811.

In 1854 a larger importation,— Lady Ayr, Red Rose, Challenge, Cherry Blossom, and Kate.[28] The two last were in the possession of Messrs. Hungerford & Brodie the year of importation, but Lady Ayr and Red Rose shortly after went to New York Mills, and Challenge has become the ancestral cow of a long line of progeny on the farm of S. D. Hungerford, at Adams, N. Y.

In 1861 arrived Dr. Hornbook, Handsome Nell, Helen Douglas in dam Lady Douglas, and Tibbie.[29] In the fall of 1862 these were still owned by Brodie, Campbell & Co., but afterwards they all were added to the New York Mills herd. At the same time a cow and a calf appear to have been imported for a Mr. Miller.

In 1864, Baldy, John Gilpin, and Tarbolton.[30] John Gilpin was retained by Mr. Brodie, the other two went to Messrs. Walcott & Campbell.

In July, 1870, arrived Lady Clyde and Lady Glasgow; the latter dropped a calf, Neptune, on the ocean, and the former a bull calf after arrival, named Lord Clyde.[31] These were all for Messrs. Walcott & Campbell.

In 1871, in the ship "Eumenides," which left Glasgow April 6, 1871, Mr. Brodie had a large number. Of these, John of Ayr and Peerless were for Mr. G. C. Bradley, of Watertown, N. Y.; Duke of Ham-

---

[28] Alb. Cult., March, 1863. A. H. B., B. 10, C. 82, 523, 726, 979. Trans. N. Y. Ag Soc. 1854. 898.
[29] Count. Gent., June 13, 1861. A. H. B., C. 149, 423, 438, 526, 791.
[30] A. H. B., B. 90. 222, 372.
[31] A. H. B., C. 1403, 1417, B. 684.

ilton, Woodville Chief, Beulah, Bessie Belle, Ayrshire Lass, Lady Ayr, Lady Rogers, Lady Mary, Lady Houston, Lady Pender, Lady Kilbirnie, Kilbirnie Maid, Ocean Belle, and Kempsey Maid were retained by Mr. Brodie. In this importation may also be included Kilbirnie Lass in her dam, and possibly a Peerless imported in the womb.[32]

It is to be regretted that Mr. Brodie has so duplicated the names of animals, either imported or owned by him, as to render somewhat difficult, in the future, the task of identifying pedigrees.

For information concerning the importations of Mr. Brodie, etc., in 1853, see under 1852.

In 1853, James Logan, of Montreal, seems to have imported the cow Buttery and the bull Baldy.[33]

The Hochelaga Agricultural Society may also have imported a bull of the same name about this time, which was used in Mr. Logan's herd, and perhaps the two animals are identical.[34]

In 1857, Mr. Logan imported the cow Stately in calf with Sir Colin, Greig in calf with Sonsie, Jean, and probably Heather Belle, Beauty, White Cherry, Red Rose, and others.[35]

In the year 1853 the Montreal Agricultural Society imported the bull Rob Roy. This importation was made through a Mr. Gilmore, and the bull appears to have been sold to the County of Leeds Society, and

---

[32] Entry Cat. N. Y. Ag. Soc. 1872 and 1873. Other references for Mr. Brodie's importations are Walcott & Campbell's Catalogue.
[33] A. H. B., C. 110, 394, 1773, and C. 62, 85, 87, 125, 186.
[34] A. H. B., C. 773, 773½, 774, 810.
[35] A H B., C. 200, B 67, 354, 385, C. 469, 463, 561, 724, 764, 289, 510, 949, 1383, B. 805, C. 86, 394, 214, 725.

afterwards to George Morton. In 1861 owned by Thomas Anthony and exhibited at New York State Fair.[36] This Society imported in all five bulls, of which the importations of 1855, 1856, and 1857 were used in the herd of Messrs. Dawes.

In this year, 1853, we find mention of the importation of a cow, Advice, by J. W. Duncombe, of Quebec.[37]

1854 An account of the 1854 importation of Mr. Brodie will be found under the date of 1852.

William Watson, of Westchester, N. Y., probably made an importation in 1854. The cow Beith, the two-year-olds Anna and Maria, were exhibited at the New York State Agricultural Fair of this year. Julia was probably imported in the womb, as was perhaps Margaret. Oswald and Sarah were possibly of this importation.[38]

There seems to have been another importation by Mr. Watson, in 1858, of the cow Kate. Taurus was imported, if at all, before 1861, and the cow Emily before 1859, and perhaps all these may be referred to this year.[39]

In 1862 Mr. Watson appears to have brought over the bull Angus.[40]

In 1868 the bull Kilbirnie.[41]

Mr. Watson selected and forwarded the animals which comprised the Sweetser importation of 1855.

---

[36] A. H. B., C. 715, 734, 737, 741, 472, B. 328.
[37] A. H. B., C. 222.
[38] A. H. B., C. 61, 76, 851, 1658; 722; B. 50, 51, C. 19, 61, 76, 168, 122, 971; 206, 698, B. 627. Trans. N. Y. Ag. Soc. 1854.
[39] A. H. B., C 1424; 329; 775½, 1125. B. 630.
[40] A. H. B., C. 851, 1014, 1168, 1424, 1474, 1544, 1658, 1833.
[41] A. H. B., B. 662; C. 1083, 1382.

In August, 1855, Mr. R. A. Alexander, of Kentucky, is said to have brought over some Ayrshires, in the ship "Olive Jordan," from Liverpool to Philadelphia.[42]

In September, 1855, Mr. Luke Sweetser, of Amherst, Mass., selected and imported through Mr. William Watson, of N. Y., four cows, Rose, Bessie, Beauty, and Tulip. Of these, Rose proved barren; and Beauty, now twenty years old, is the property of the Massachusetts Agricultural College.

During this same year, 1855, Mr. John Dods, of Montreal, imported the cow Ann. He seems to have made other importations as follows: Previous to 1859, Pailey; to 1860, Baldy and Bonnie Scott; to 1862, Jane; to 1863, Annie and Del, or Delavan; to 1864, Blackie and Cherry 1st, and previous to 1866, Lord Eglinton and Maggie.[43]

In 1856, James Gibb, of Canada, imported a bull, Major, and a cow, Fairy.[44]

Mr. Wm. Chambers, of St. Laurient, near Montreal, is said to have imported a cow in 1857,[45] which perhaps was named Rose.[46]

For particulars of Mr. James Logan's importation of this year, see under date of 1853.

For the importation of the Massachusetts Society for Promotion of Agriculture of this date, see under

---

[42] Count. Gent., Sept. 13, 1855.
[43] A. H. B., C. 230; C. 1279, 1698; C. 998, 1279, 1698; B. 70, C. 3, 121, 142, B. 643; C. 1279, 1405; B. 539, 482, C. 1803, 1881; B. 879, C. 1422, 1488; C. 992, 1308; C. 1716; B. 574, 771.
[44] A. H. B., C. 1156, 1433, 1439½. C. C. Abbott's Sale Cat. 1870. J. L. Gibb's Cat. 1870.
[45] A. H. B., B 247, C. 705.
[46] A. H. B., C. 512, 540, 854.

the year 1837. For that of Mr. Watson, see under date of 1854.

1858 In 1858, Messrs. Dawes, of Lachine, P. Q., commenced their series of importations with the bull Rob Roy.[47]

In 1860 they imported a bull, Prince, and the cow Queen of Scots, in calf with Duchess.[48]

In 1868, the bull Samson.[49]

In 1869, the cows Picture, in calf with Lily, and Portrait.[50]

In 1870, the cows Empress, Marchioness, Miss Henry, and Miss Kennedy. These all were with calf, and produced the heifer calves Medora, Basay, a third heifer, and a bull.[51]

In 1871, the bull Nicholas and the females Snowdrop, Turnlow, Beauty, Pride of Avon, Lady Bird, and Drumbowie. With these were brought over five heifer calves in their dams.[52]

In 1858, Mr. H. H. Peters, of Southborough, Mass., made his first importation. He authorized Mr. Sanford Howard, who was going to Scotland, to procure stock for the Massachusetts Society for Promotion of Agriculture, to purchase for him four Ayrshire heifers,[53] which are those which were named Jean Armour, Miss Morton, Miss Miller, in calf with Miller 2d, and Miss Betty.

In 1859, being well pleased with the cattle of the

---

[47] A. H. B., C. 413, 492, 975, 1043, 1132, 1139, 1267, 1613; B. 404.
[48] A. H. B., B. 354, 425, 589, 821, 876½; C. 910, 975, 986, 1007, 1132, 1185, etc.
[49] A. H. B., C. 896, 939, 1033, 1036, 1070, 1159, 1229, etc.
[50] A. H. B., C. 1720, 1457, 1733.
[51] A. H. B., C. 1185; 1535; 1598; 1599; 1572; 867.
[52] A. H. B., B. 755; C. 1843; 1908; 876; 1737; 1396; 1088.
[53] H. H. Peters' Cat. U. S. Dept. Ag. Rept. 1863, p. 198.

previous year, he engaged Mr. Howard to go to Scotland expressly on his account to select and purchase. In August, 1859, there arrived two bulls, Eglinton and King Coil, and twenty-one cows and heifers. Ada, who proved barren, Alice, Beauty, Brenda in calf with Brenda 2d, Corslet, Duchess 2d, Flora, Jane, Maggie, Mary 3d, Minna, Miss Drew, Mistress 2d, Nannie, Pink in calf with Oswald, Queen 2d in calf with Empress, Queen 3d, Rosa, Ruth, Susan in calf with Susan 2d, and Young Merryton in calf with Merryton 3d.[54]

While Mr. Howard was in Scotland he appears to have selected cattle for others. Such was the case with the cow Effie, imported by Mr. Rufus Carter, of Worcester, in 1858;[55] and probably the cow Margery, imported by Mr. J. S. Cabot, of Salem, in 1858;[56] the cow Jessie, imported by Dr. George B. Loring, of Salem, the same year;[57] and the importations of Mr. Lyman were of his selection. Geo. W. Lyman, of Waltham, appears to have received the bull Comet, possibly two cows, one to calve with the bull calf Scotland.[58]

In 1858 also, Mr. John Brooks, of Princeton, Mass., imported a bull, Dr. Hornbrook.[59]

Capt. Peel, of Canada, is said to have brought over in this year a bull, Roxborough, and a cow.[60]

The importation of Mr. C. M. Pond, of Hartford, Conn., made this year, appears to have consisted of

---

[54] H H. Peters' Cat.
[55] A. H. B., C. 858.
[56] A. H. B , C. 135.
[57] A. H. B., C. 107.
[58] A. H. B., B. 124, 131, 132, 133, 316. C. 240, 374, 419, 437, 476, 606, 742, 782, 805, 1746; C. 132, 133; C. 357.
[59] A. H. B., B. 16. H. H. Peters' Cat.
[60] A. H. B., C. 1569, 1245.

the yearling bull Robert Burns, and the two-year-olds Jennie, Jessie, and Rose of Brown Hill.

His importation of 1859 was Ayrshire Lassie, Lily of Smithfield in calf with Cardigan, and a heifer, Bella.[61]

1859 For an account of the importation of Mr. H. H. Peters for this year, see under date of 1858, as well as that of Mr. Pond.

In 1859, a Robert Gray, of Fredericton, New Brunswick, is said to have imported a bull, Geordie.[62]

Mr. Gardner Brewer, of Boston, imported a cow, Flora, in calf with Robert Fulton, during this year.[63]

Three two-year-old heifers, Fanny Ellsler, Florena, and Lady Ellen, were imported in 1859 by Mr. H. E. Day, of Hartford, Conn.[64]

Mr. Eben S. Poor, of South Danvers, Mass., imported this year two cows, Lily in calf with Duke, and Rosa in calf with Bessie,[65] and possibly a bull.[66]

About the year 1859, or previously, there seems to have been imported the animals known as the Cuthbert bull and cow, by Mrs. Cuthbert, of Lanoraie, P. Q.; or perhaps this importation is the same with that of Mr. Cuthbert, of Berthier, who seems to be credited with some about the same time. One of these animals is apparently Maggie Lauder.[67]

1860 The importation of Messrs. Dawes, for 1860, has been noticed under date of 1858.

---

[61] A. H. B., B. 57; C. 98, 111, 190, 4, 130; B. 265; C. 409, 961, 1651.
[62] A. H. B., B. 23.   [64] A. H. B., C. 72, 78, 123.
[63] A. H. B., C. 388, B. 318.   [65] A. H. B., B. 356.
[66] Count. Gent., Feb. 2, 1861. Alb. Cult., Apr. 1860, p. 130.
[67] A. H. B., C. 1026, 1020, 601, 1027, etc. Abbott's Sale Cat. 1873, A. H. B., B. 197, 635, 539; C. 1034, 992, 1027, 1405, 1624, 581. Whitney's Cat. 1871.

In 1860, Mr. John Chambers, residing near Montreal, appears to have imported a three-year-old heifer, Strawberry.[68]

The account of the importation of Brodie, etc., for this year, will be found under date of 1852.

In 1861, S. Beattie, of Canada, imported in ship "Helen Douglas," at the port of Quebec, an Ayrshire cow.[69]

1861

Mr. Beattie appears to have imported a cow, Mountain Maid, which possibly is the animal referred to above, and at a later date a bull, Carrick Farmer.[70]

It was in this year that a Mr. Miller is said to have imported a cow and a calf on the ship "Helen Douglas," at Quebec.[70] These importations being on the same vessel with Brodie and Campbell's, were possibly of their selection.

The importation of Mr. Watson for 1861 has been noticed under date of 1854.

In the spring of 1863, J. M. Browning, of Beauharnois, P. Q., seems to have imported the cow Effie in calf with Daisy.[71] It is possible that the bull Marquis, said to have been brought over by the Beauharnois Agricultural Society, was imported at this time.[72]

1863

The importation of Brodie & Co. for 1864 has already been noticed under date of 1852.

1864

In June, 1864, J. L. Gibb, Esq., of Compton, P. Q., commenced his series of importations with the

---

[68] A. H. B., C. 1866.
[69] Count. Gent., June 13, 1861.
[70] A. H. B., B. 883; C. 825, 845, 894, 1092, 1098, 1267.
[71] A. H. B., C. 1104, 1046.
[72] A. H. B., B. 709.

five-year-old cows Quess and Lily, the yearling heifer Gypsey, and the bull Marquis.[73]

In August, 1868, he brought over the two-year-olds Princess Alice and Princess Royal,[74] with Florence and Hebe in their wombs, and the yearling bull Mars.

In 1870 two importations. The one in June consisted of the five-year-old Annie; the four-year-olds, Medora in calf with Medora 2d; Lina in calf with Merryton Lass, and Flora; the three-year-old Emma, who dropped the heifer Atlanta on shipboard; the two-year-olds Lily 2d, Park 2d, Rossie, Roughhead 2d, Blackhouse 2d, Gartnoad 2d, and Alice.[75]

In September, 1870, the yearling bull Glenluce, and the four-year-old Lady Avondale, in calf with Lord Avondale; the two-year-old Mary in calf with Earl of Lorne, and Beauty; the yearling heifers Blooming Daisy, Mary Belle, Miss Meikle, Heather Belle, and Lass o' Gowrie.[76]

In September, 1871, the two-year heifers Verbena and Crocus.

In August, 1873, in the steamship "Hibernian," at Quebec, Mr. Gibb imported the cows Clarinda, Heather Bloom, and Heather Bell, Derby, and the two-year heifer May Morn.[77]

1865 Mr. Thomas Miller, of Brushland, Delaware Co., N. Y., made his importation of the cow Daisy in

---

[73] A. H. B., C. 1769, 1234, 1235, 1237.
[74] A. H. B., C. 1749, 1750, 1751.
[75] A. H. B., C. 348, 1573, 1574, 1578, 1897, 1200, 1130, 856, 1464, 1702, 1810, 1813, 935, 1222, 831.
[76] A. H. B., C. 144. B. 682, 537. Count. Gent., Nov. 3, 1870.
[77] Count. Gent., Aug. 28, 1873.

calf with Favorite, and the bull Duke of Hamilton, 1865 in the ship "John Phyfe," which arrived at New York, May 1, 1865.

At the same time was imported Rosy or Scotch Rosy, who was afterwards purchased by Mr. Miller.

For the details of James Logan's importation of this year, see under 1853.

For the details of J. L. Gibb's importation for 1868, see under date of 1864, and that of Dawes 1868 under 1858.

For that of Mr. Watson, see under date of 1854.

In November of this year, per steamship "Java," at the port of New York, Mr. G. D. Cragin, of Rye, imported the cows Rowena in calf with Hero, Edith, Duchess of Hamilton, Queen Bess, Queen Mary, and a bull, Duke of Hamilton.[78]

In the barque "Melbourne" from Ardrossan, in December, 1868, Mr. H. W. Tilton, of Walpole, Mass., received a pair of Ayrshire cattle, Earl of Holderness and Lady Harmonie.[79]

William Semper, of Allegheny City, Penn., is stated to have imported Clydesdale and Lily in 1868.[80]

In 1868, or thereabout, an importation of a bull, Robbie Burns, is claimed for Thomas Irving, of Rockfield, near Montreal.[81]

For the importation of Messrs. Dawes in 1869, see 1869 under date of 1858.

---

[78] Count. Gent., Oct. 7, 1869. A. H. B., B. 612, C. 1099, 1759, 1762.
[79] Count. Gent., Dec. 24, 1868. A H. B., B 536; C. 1421.
[80] A H B., B. 730.
[81] A. H. B., B. 804.

1869   In July, 1869, arrived the first of Mr. N. S. Whitney's importations at Montreal, the yearling bull Jock, and the two-year-old Bessie Bell.[82]

In June, 1870, Mr. Whitney received, per ship "Geneva," the four-year cow Clara, who dropped on shipboard the bull calf Pride of Geneva; Kelso, three years old, also dropped a bull calf, Sailor, while on shipboard; Maggie, in calf with Marquis of Bute, and the two-year-olds Netty and Dow 2d.[83]

On the 14th of September, 1870, arrived Barrochan Maid, and Bonnie Lassie in calf with Bonnie Lassie 2d.[83]

In 1871 still another importation by Mr. Whitney. Eleven head arrived in September, in the ship "Abeona." These were Daisy, in calf with the bull Fleetwood; Rosie, who dropped a bull calf, Neptune, while on the water; Flora in calf with Flora 2d; Stately, and Beauty who dropped Beauty 2d, on shipboard.[84]

1869   In 1869, Mr. M. P. Cochrane, of Canada, imported two Ayrshire heifers, probably Lady of the Lake and Maggie.[85]

In 1871, or thereabout, Mr. Cochrane seems to have imported the bull Champion, and the cows Daisy, Cocksey, Village Maid, and Mary Gray.[86]

On the 9th December, 1869, arrived on the steamship "Nova Scotian," at Portland, the importation of

---

[82] N. S. Whitney's Cat., Jan. 1870. A. H. B., B. 641, C. 926.
[83] N. S. Whitney's Cat., Jan. 1871.
[84] N. S. Whitney's Cat., May, 1872. Count. Gent., Oct. 5, 1871.
[85] Count. Gent., Sept. 2, 1869. Entry Cat., N. Y. Ag. Soc. Fair, 1872.
[86] Entry Cat., N. Y. State Ag. Soc. Fair, 1872.

Sturtevant Brothers, Waushakum Farm, South Framingham, Mass. It consisted of eight cows, Edna in calf with Glengarry; Ozora in calf with Ocena; Drusilla in calf with Domine, whose name was afterwards changed to Shotto-Douglas; Queen of Ayr, in calf with bull Mains; Ops in calf with Eos; Twinney in calf with Euona, whose name was afterwards changed to Alice Brand; Mona in calf with Banquo; and Selena, in calf with Asmodeus.

The importations of Mr. Gibb for this year have already been noticed under date of 1864; those of Mr. Whitney under date of 1869; that of Messrs. Dawes under date of 1858; and under date of 1852 that of Brodie & Co.

In July, 1870, Mr. William Gibson, of Morrisburg, P. O., imported six cows and a bull in ship "Thomas Hamlin." These appear to have been Ranting Robin, Jennie, Rosa, Maggie, Edith, Princess in calf with Thomas Hamlin, and a Maggie, in calf with Robert Burns.[87]

1870

Maggie and Edith appear to have gone to J. T. Rutherford, Waddington, N. Y., and are referred to as of his importation.[88]

In July, 1870, per ship "Abeona," Mr. J. J. C. Abbott, of St. Anne's, near Montreal, seems to have imported the cow Lilias and the bull Sir Roger. Young Primrose, Young Mary, Young Dandy, Young Beauty, and the bull Yellow-Haired Laddie may also be credited to this arrival. Mr. Abbott had also in

---

[87] A. H. B., B 791½; C. 1316, 1313, 1781, 1519, 1097, 1746½; R. 888.
[88] A. H. B., C. 1519, 1097.

1870 his possession, and probably imported Alison, Annabel, and Abeona, imported in dam Annabel.[89]

It is possible that Darling 3d, and Geneva with her calf Sir Hugh, were imported by Mr. Abbott, although we have found references to Mr. C. C. Abbott as their importer in the ship "Abeona."[90]

In 1873 Mr. Abbott again imports,—this time two cows, Viola and Elsie.[91]

In 1870, William H. T. Hughes, agent for L. P. Fowler, of England, an importer by business, introduced eight cows in calf per ship "Rhine," in his first invoice, and nine cows and a bull per ship "Plymouth Rock," from London, in his second.

Of his first importation were the cows Betty Burke and Scotia.

Of his second importation the cows Beauty, Cozie, Buttercup, Cowslip, Ayrshire, and Ayrshire Bell.

In July, 1870, on ship "Thomas Hamlin," at Montreal, Mr. J. H. Morgan, of Ogdensburg, N. Y., brought over his first importation, the bull Habbie Simpson, and the cows Model of Perfection, Minnie, and Nancy. Minnie passed into the possession of D. Magone, Jr., and Nancy, of Z. B. Bridges, Esq., both of Ogdensburg; and Model of Perfection at a later date was sold to Sturtevant Bros. for $1,000, the highest price known to have been paid for an animal of this breed.

In April, 1871, Mr. Morgan made his second importation in the ship "Eumenides." It consisted of

---

[89] A. H. B.. C. 1455; B. 868. Count. Gent., July 14, 1870. C. 836, 844, 822.
[90] C. C Abbott's Sale Cat
[91] Count. Gent., July 31, 1873. Abbott's Sale Cat.

the bull Adino, and the cows Annie, Bessie in calf 1870 with Sea King, and Georgie, in calf with the heifer Sea Bird. Annie was transferred to D. Magone, Jr., while Georgie and Sea Bird went to Sturtevant Bros.

Having perused the entire correspondence between Mr. Morgan and his agent abroad, we think there is little doubt but that these importations were of the best stock in Scotland; and to the rivalry induced by this first importation may be ascribed the exceptional quality of the importations of this and the following years.

In September, 1870, Thomas Thompson & Son, of Williamsburg, P. O., brought over an importation of eight, selected personally. These were the two-year-old bull Crown Prince, and the yearling bull Highland Chief, the cow Diamond in calf with Hansom, and the heifers, Annie in calf with Queen of Beauty, Rassie 2d in calf with Rose of Carron, Rassie 3d in calf with Queen of Scots, and the calves Duchess and Bonnie Jean.[92]

In October, 1870, Mr. Thomas Paterson, of Gouverneur, N. Y., imported per steamship "Sweden," at Quebec, a pair of yearlings, Lord Raglan and Beauty.

In the same vessel, at the same time, Mr. Andrew Allan, of Montreal, brought over the two-year-old bull Boydstone, and the cows Susan, Fleckie, Rogers, and Kate.

In June, 1871, per steamships "European" and

---

[92] Thompson's Catalogue, 1871.

"Nova Scotian," the cows Belle of Straven, Barbara Allan, Fairy Queen, Straven Maid, and Straven Queen.

In September, 1871, per ship "Abeona," the bull Conquer.

1871 For a notice of the importation of Mr. Allan for 1871, see under date of 1870; for that of Mr. Cochrane, see under date of 1869; of Mr. Morgan, under date of 1870; of Mr. Whitney, under date of 1869; of Messrs. Dawes, under date of 1858; of Brodie & Co., under date of 1852; and Gibbs, under 1864.

July 31, 1871, per ship "Gluco," at Montreal, Mr. James McNee imported the yearling bull Robert Burns, and the two-year-old heifers, Highland Mary and Ayrshire Maid.

October 31, 1871, Mr. Charles H. Peckham, of Providence, R. I., imported, in barque "J. B. Duffries," three heifers, Highland Maid, Village Belle, and Sally.

1873 For the importation of Mr. Abbott for 1873, see under date of 1870. For that of Mr. Gibb, see under date of 1864.

In October, 1873, Irving Moyer,[93] of Fort Plain, N. Y., imported the two-year-old bull Sir John Moore, and the cows Lady Martha and Lady Mariam. The calves Heather Jock and Damsel were imported in these cows.

In a list by themselves we place those importations which we are unable to identify by a certain date.

---

[93] Count. Gent., Oct. 15, 1874, p. 666.

IMPORTATIONS.

Mr. R. S. Griswold is said to have imported a bull, Juba, and cow, Whitey, some time previous to 1849.[94]

Mr. Nicholas Biddle, of Philadelphia, seems to have made an importation previous to the year 1850.

Between 1850 and 1854, Mr. Peter Lawson, of Lowell, appears to have imported a bull, McDuff.[95]

A Capt. Smith is credited with having imported a cow, Cherry, which must have been within the decade, 1850 and 1860.[96]

Sir George Simpson is said to have imported a cow, Lady Simpson, somewhere about this time.[97]

Previous to 1855 Col. Beatson appears to have imported a cow, Lady Betty.[98]

Previous to 1858 the Hochelaga Agricultural Society, of Montreal, imported the bull Bauldie, and afterwards the bull Buchanan, and another without name.[99]

Previous to 1858 the Montreal Agricultural Society is credited with having imported a bull, Bauldie, and previous to 1860 a cow, Queen of Scots.[100]

Somewhere near 1860, Mr. Thomas Richardson, of West Farms, New York, appears to have imported a pair, Eric and Norna, and possibly Norval, in dam.[101]

A cow named Sally appears to have been imported by a Mr. Hutchinson previous to 1860.[102]

---

[94] A. H. B., C. 105.
[95] A. H. B, C. 100, 127, 141.
[96] A. H. B., C. 593½, 705; B. 394.
[99] A. H. B., C. 773½, 774, 810, 1460. King's Cat. 1872.
[100] A. H. B., C. 447, 1132, 1446, 1729; B. 72.
[101] A. H. B., B. 630; C. 673½, 676, 775½, 1125, 1147, 1021, 1050, 1691, 1692.
[102] A. H. B., B. 785; C. 893, 998.
[96] A. H. B., C. 1808, 1940.
[97] C. C. Abbott's Sale Cat. 1870.

A bull was imported by a Mr. Burstall, of Quebec, before 1860.

Mr. R D. Shepherd, of Va., is said to have imported a bull, Brutus, previous to 1859.[103]

Mr. Charles Jones, of Brockville, Canada, is credited with the importation of a cow, Bonnie Lass.[104]

J. Gilmore, of Quebec, is credited with having imported a cow named Buttercup previous to 1865. He is also said to have imported Rob Roy for the Montreal Agricultural Society. Is the Rob Roy of the 1853 importation of this Society the same bull?[105]

Before 1864 the bull Ayr 2d was imported by Mr. Perreault of the "Canada Agriculturist."

Some time in 186– it is claimed that Mr. Wm. E. Lockwood, of Penn., imported a pair, Zero and Kate.[106]

About 1867, Mr. W. Rodden, of Montreal, P. Q., appears to have imported Scotch Mary, Snow Drop, and Nancy of Ayr.[107]

Some time later than 1867, Mr. J. Laurie, of Scarboro', P. O., is said to have imported Avondale Farmer and Dutchy.[108]

Mr. Patrick B. Wright, of Coburg, Canada, is said to have imported Young Percy and Buttercup before 1868.[109]

---

[103] A. H. B., C. 99.   [104] A. H. B., C. 1941.
[105] A. H. B., C. 706, 737.
[106] Entry Cat. N. Y. State Ag. Soc. Fair, 1872.
[107] A. H. B., C. 1644, 1831, 1842.
[108] A. H B , B. 422, C. 1094.
[109] Thomas Thompson & Son's Cat. 1871.

Between 1870 and 1872, Mr. Simon Beattie apparently imported the bull Young Prince and heifer Straven Callen.[110]

The Hamilton Agricultural Society of Canada seems to have imported a bull through Hon. Michael Cameron; this bull afterwards came into the possession of a Col. Astley.[111]

---

[110] Entry Cat. N. Y. State Ag. Soc. 1872, p. 18.
[111] C. C. Abbott's Sale Cat.

PEDIGREE.

Those who are at the expense of introducing a foreign breed of cattle, are generally desirous of preserving it untainted from interbreeding with the cattle by which they are surrounded. They desire also to preserve some memoranda of each individual of the foreign breed, both for present use and future reference. They are thus enabled to breed them more understandingly, for they know whether the animals mated are akin or not, as also whether a particular animal has originated from ancestry of a particular or desirable type. If there be one herd only of the new stock in the country, the owner must have notes either written, or preserved in memory, or the stock is likely to deteriorate. If he trusts to memory, and upon his death the stock passes into the hands of strangers, without further knowledge of the animals than comes of seeing them, much of the value of the animals have departed with the demise of their owner. When the new breed is somewhat disseminated, and there are many herds, breeders find it advisable to seek occasionally an interchange of blood. But no breeder will do this without the fullest assurance of the stock he seeks being pure bred, and without knowing, if possible, from what parentage the animal has come.

The breeder's object being first to produce good animals, and second to secure remuneration from their sale, it is important to have a regular and systematic plan for making his efforts known both to those engaged with the new breed, and others who may be desirous of adopting it. By such a plan, not only will higher prices be realized, but the breed will become more disseminated.

To effect these several objects, — the preservation of a breed in its purity, the maintenance of the excellence already attained, the securement of a progressive improvement, the advertising and thus facilitating sales, the guaranteeing of the expected purchasers against fraud to a large extent, — all those interested in a common breed unite in the support of what is known as a herd book.

This is, or should be, a printed volume. It should contain the name, with a number attached, of each animal of the breed imported, from whence imported, by whom bred, for and by whom imported, ship, port of entry, and date of arrival, — a description of the animal sufficiently minute for identification, with the age and sex.

Starting with imported animals as a foundation stock, their descendants alone should be entered, with description of each, date of birth, by whom bred, by whom owned, and names and numbers of sires and dams to importation.

The value of a herd book, in every case, depends upon its fulness, completeness, and reliability. If it is found easy to enter a grade animal in a book designed only for the imported stock, and the progeny of imported stock, it is far from impossible that some persons will take this cheap course of fraud, herald their grade animals as pure, and obtain for them a place in this choice company. The need, then, is manifest of adopting some rigid conditions, conformable to certain principles of utility, and the abiding by them persistently. For not only does the fraud of entering grade animals produce a lowering of the quality of the breed, but by producing an abundance of low-cost stock, prevents the more careful and exemplary breeder from selling a stock, costing higher and of more value, at remunerative prices. A herd book which will allow of this, acts to discourage the honorable breeder, and tends to drive him from the field in despair.

The incorrectness of a herd book, known at first perhaps only to a few persons, imposes obstacles for a while upon the many; but afterwards, by becoming known to many, induces a want of confidence in pedigrees, faith in which is so conducive to success in breeding stock to a high degree of excellence.

It is a cause for regret that the earlier importers of Ayrshire cattle did not foresee the advantages that were to be derived from a herd book, and the disadvantages that would attend its absence. When the attempt was made at the late date of 1863, there were herds of cattle in the country, thought by their owners to be too valuable to be excluded from such a work, — cattle undoubtedly Ayrshire, but of ancestry so ill-defined and uncertain, that their admission to registration precluded the rejection of animals far more objectionable. The volume of this year records the names of 79 males and 216 females. Of the males, the number stated to have been imported is 11; the number whose ancestry is traced unbroken to importation is 50; others, 18. Of the females, the number entered as imported is 57; traced unbroken to importation, 109; others, 50. There are thus 68 animals recorded "on the assurances of well-known breeders that the animals in question are thoroughbred Ayrshires." These assurances are not founded upon definite information as to their breeding; they may be true, but there appears to be no evidence presented that the assurances are anything more than selfish opinions.

With the appearance of the second volume in 1868, the number of Ayrshires, the pedigrees of which are presented complete or unbroken, is near 1,300. Of the bulls in Volumes I and II, about 280 appear to be traced to importation; about 120 are not so traced. Of the females traced to importation there are about 530; not so traced, about 228.

Each of these volumes bears the title, "Herd Record of the Association of Breeders of Thoroughbred Stock, Ayrshire." The second has a recognized editor, J. N. Bagg, of West Springfield, Mass.

In 1871, Volume III appeared, with Mr. Bagg for editor, but with a new title, "The American and Canadian Ayrshire Herd Record." The Canadian portion is of Canadian editorship, over which it appears the American editor exercised no supervision.

In this volume the number of bulls registered is brought to number 931, of females to number 1,951. We have thus far recorded 59 imported bulls, and 192 imported cows. Of the total pedigrees to date, 1,354 are traced to importation, and 1,321 are not traced to importation.

We will place the result of our analysis of the "Bagg" Herd Book in the form of a table.

Number of animals recorded: —

Vol. I.........	79 Bulls,	216 Cows.	Total 295
" II.........	342 "	617 "	" 959
" III.......	527 "	1,145 "	" 1,672
	948	1,978	2,926

Number of imported animals recorded: —

Vol. I	11 Bulls,	57 Cows.	Total........ 68
" II.........	10 "	18 "	" 28
" III.......	38 "	117 "	" 155
	59	192	251

Number of animals recorded properly: —

Vol. I.........	50 Bulls,	109 Cows.	Total........ 159
" II	230 "	421 "	" 651
" III.......	202 "	342 "	" 544
	482	872	1,354

Number of animals improperly recorded: —

Vol. I	18 Bulls,	50 Cows.	Total........ 68
" II.........	102 "	178 "	" 280
" III.......	287 "	686 "	" 973
	407	914	1,321

Percentage of poor pedigrees[1] in Vol. I, 23
" " " " II, 29
" " " " III, 58

Average for all, 45 per ct.

---

[1] By poor pedigrees is meant, Recorded improperly. Some few classed here are correct, but not shown so by the record; others present no claims for correctness other than the fact of admission to the record.

In the index to Volume I, we find the names of 129 owners; in Volume II, this number has increased to 206; in Volume III, to 322.

The average number of animals recorded to each name was 2.45 in Volume I, 4.65 in Volume II, and 5.06 in Volume III.

The distribution by States in Volume III is as follows: —

Alabama	1	New Jersey	13
Connecticut	22	New York	76
Florida	1	Ohio	11
Illinois	4	Pennsylvania	8
Indiana	1	Rhode Island	13
Iowa	1	Vermont	20
Kansas	1	Virginia	1
Maine	7	Wisconsin	1
Massachusetts	90	New Brunswick	2
Michigan	6	Canada	30
Mississippi	1		
Missouri	1	Total	322 owners.
New Hampshire	11		

It seems the less necessary to enter upon a full explanation of the attempt to furnish a herd record for the breed, from the circumstance that the work is wholly unsatisfactory to breeders and purchasers, who have examined it, and the volumes are only awaiting the action of some responsible person, or association, to start a register upon sounder principles, when the present work will be ignored.

After this arraignment of the present Herd Book, it seems right that we should point out a few of the errors, in order to justify ourselves in the position we have taken.

Misprints. — The wrong number occurs to Sachem Chief, in C. 474; Star of the North, in B. 759, should be 876½. After 321 in C. 849, should be 221; Jethro in Diva 1081 should be 638 and not 628, etc. Andover, not Adams, in Moss Rose 1620; Maggie 161 in Lotty 1495 should be Maggie 1564; Daisy 1043 in C. 1175 should be 1045, etc.

There are frequent errors in names, as Allan for Allen, Graig for Greig, in 1333; in names of places, as Stathaver instead of Strathaven, in Geordie 573, and Lannarkshire instead of Lanarkshire, etc.

Omission of Numbers, — as 250, 473, 515; also, 488, 489, 503, 505, 521, 565, 744, 767, 1332, etc. etc.

Carelessness. — Sea Bird 847½, and Vashti 901, among the bulls, should be recorded among the females. Robert Bruce 314 and 808 are the same animal; as are also, in all probability, John Gilpin 652 and 653; Robert Burns 810 and 811; Thomas Hamlin 888 and 889; Belle 256 and 899; Bonnie Jean 289 and 969; Lady Ayr 523 and 1394; Maggie 1319 and 1521, etc. etc.

Pride of Geneva is entered among bulls, No. 779, and also among the cows, No. 1739.

Lady Bruce 1397 has no pedigree given whatsoever. The same remark applies to Lady Prentice 124, and others.

Lack of Editorial Supervision. — Lady of the Lake, in Prince Arthur 783, was calved same year as her granddam.

Star of the North 876½ calved May 7, 1866; dam, Duchess 1090.

Lanark 670 calved March 24, 1866, dam, Duchess 1090.

How could Victor 904 be bred by Thomas Dawes & Son, Lachine, P. Q., when her dam was a thoroughbred cow in Scotland?

Tam o' Shanter, in Buttercup 970, was imported from New Brunswick, not from Scotland.

Dew Drop 338 and Dew Drop 1062 are the same animal, with a difference of ten days in birth.

Heather Bell 1267 and 1268 are the same animal. In one entry the dam Florence is by an imported bull, in the second entry she is imported.

Spotty (148) in Juno 1366 is not by McKenzie, imported.

Maggie Morton 1528 and Effie Morton 364 have same parents, — calved in 1868, one in February, the other in January.

Sailor 835 is credited with Kelso 1385 for a mother. Kelso was calved 1870; Sailor the same year.

Medora 2d 1574 was calved in July on board ship; her dam Medora 1573 was imported in June, 1870.

Ayrshire Mary 864 could not be "bred by Mr. Rodden," when sired by a bull in Scotland.

In some cases we have very serious errors. For example: John o' Groat, dam and sire both imported by R. S. Colt, as we are informed under Jock 643. Under St. Andrew 874 the sire is said to have been imported by Capt. Nye. Under Young America 925 the dam comes from an imported bull and cow, imported by Capt. Nye. In still another place, under Fairie 1153, we have an entirely different account. Under Kate 1369 John o' Groat is said to be "imported." Under Lady Geraldine 1416 this name again appears.

Another class of errors is when the same cow has two calves at periods much closer together than usual. For instance: Lady Mary 536 and Sir Colin Jr. 68 were both calved by Heather Bell 86, — the one in March, the other in April, 1860. They are recorded as half-sisters.

Cornelia 35, Belle 256, Belle 899, Logan 45, were all born from Heather Bell between April 8 and February 14, 1862.

Jessie 6th 497 and Dick 147 were also born of different fathers, same mothers, and yet within five days of each other.

Suegat 363 and Flora 386 are hardly better off.

Cowslip 2d, Lassie, and Cowslip 3d were all born from the same mother but different fathers, in the same year, 1864.

Highland Lassie 2d and Daddy Auld were born within a month of each other, but had the same father. Peverel and Daisie afford another instance, as well as Flora Temple and Rosa, Highland Mary and Queen Mary (possibly twins), Lady Gowan and Kilburn, and others.

In truth, the errors in this Herd Book are too numerous to mention. We do not think a half-dozen pages can be selected from Volume III which shall be entirely free from error. Opening at random, we examine page 72 : but one correctly recorded pedigree. (Sea Bird 847½ is a heifer.) The following page is no better, nor is page 75 an improvement. Page 76, again, contains six poor to one good pedigree, etc. etc. Are the cows any better recorded in this volume? We examine page 110, and do not find a single perfect pedigree. Page 111 is scarcely better: but one pedigree, outside of the imported animals, that can be pronounced good. A little search shows page 136 without a single perfect pedigree, and we find many others with but one each; and not yet have we found one perfect page in this volume.[1]

Having now briefly noticed the deficiencies of the American and Canadian Ayrshire Herd Book, a few reflections concerning the significance of pedigree may be in place.

A pedigree is more or less complete according as the animals are traced backward through several generations, with or without omission of any of the ancestry. If we know the earlier parents, and are unable to trace the connective link that ties them in relationship to the animal under consideration, then the missing links are so much out of our knowledge of the animal; we come just so far short of acquaintance with the antecedents of our animal. The importance of knowledge of the antecedents of animals from which we desire a succession, depends on the circumstance, in great part, that all animals are what they

---

[1] Since these strictures were penned, the authors of the present book have inaugurated, at the request of breeders, the "North American Ayrshire Register," which is already far advanced on an apparently successful course.

are in form, in mind, in capacity for useful services to man, because particular individuals rather than others are related to them. The mating of male and female not only ensures offspring, but offspring impressed with the individual stamp of the parentage, more or less disguised. A change of mating is followed by a changed character of the fruit, and each parent contributes to form the general mould in which the offspring is cast.

The influence of near ancestry is commonly more obvious than of ancestry lying at three or four or a dozen removes. Peculiarities we are apt to ascribe to the moulding force of near kindred; but we should not forget that this moulding force passes from generation to generation, and that the animal before us is the outcome of successive steps, of which neither the sequence, nor the character, could have been different without occasioning an animal different in some particulars.

Improvement is not readily fixed in a family by two or three or a half-dozen successive judicious matings. No existing breed of cattle, of marked value, is less than a hundred years old, though it is a little less time since the value of the stock became widely recognized and the record of marked improvement begins. We know what value is attached to Short-horn cattle whose lineage can be traced to famous animals living in the latter half of the last century. The best Devons descend from animals of local fame living in the early part of the present century. The fountain-head of the breed was with one family, the Quartlys, who gave them reputation, and who have kept the lead since; and in 1850 two neighbors had kept up this breed in their families for more than a hundred years.[2] A large number of the Devon breed in America have traceable lineage to well-known animals, prize-takers wherever shown. The history of this stock bespeaks the value of antiquity of pedigree, but it is the same with every breed that has had as full opportunities of development. Commonly the most valued Ayrshires in Scotland have most length of traceable pedigree. It is not always the last mating, but often a mating several removes back, that the Scotch breeders refer to; and he who remembers this and subsequent matings, is likely to be among the winners at the fairs so much in favor with them; and if one looks over the premium lists for many years, the chances are he will find that of the careful breeders, the oldest win most frequently when the competition lies between animals bred by the owners.

These statements and considerations are calculated to impress upon us, appreciation of the importance of a knowledge of pedigree. It matters little whether we have a mixed breed, common cattle, or the thorough-bred. The same need of regarding the ancestry exists. If we would have much assurance of what nature of cattle we may have from our breeding herd ten years hence, we require to know much of the ancestry of the animals that formed the herd for the ten years past. For we rely on the past, with

---

[2] Jour. R. A. S. of Eng. 1850, p. 681.

the present, to guide us to the future. To desire to build upon the present alone, regardless of the past, is as wise as it would be for the architect to build his stone house without attending to the nature of the foundation upon which it is placed.

We think the lack of knowledge of the ancestry of choice animals imported from abroad, occasions the frequent observation, that their progeny is inferior to themselves. The young, with which they often come laden, may be superior to the mother, and not uncommonly are of fine quality, but the produce of the American breeding is a disappointment. The valuable imported animal is thus shorn of a part of his value when put into strangers' hands. It is therefore rarely the case that the much-praised foreign animal quite fulfils expectations, when moved from his native neighborhood and put to breeding. At home, acquaintance with its pedigree, the animals that enter into it, with their merits, defects, and tendency to the cropping out of particular traits, is put to practical use.

To realize, in its fulness, the idea of a pedigree, would be to bring into array before us the living animals, and sun-portraits of the deceased ancestry. To realize the idea of a Herd Book in its fulness, would require that there be introduced in the volume the sun-portrait of every animal named; and the naming of all animals without any omission for many generations. It is well to carry this ideal in our minds, and, rejecting what is manifestly impracticable, realize all we can.

IMPORTED PRIZE AYRSHIRES.

PRIZE-TAKING IN SCOTLAND A GUARANTEE OF AUTHENTICITY OF BREED.

Abbott's	Alison,	Elsie,	Viola,
	Darling,	Lilias,	Yellow-haired Laddie.
Allan's	Barbara Allan,	Boydstone,	Conquer,
	Belle of Straven.		
Brodie's	Lady Douglas,	Lady Kilburnie.	
Gibb's	Annie,	Lily 2d,	Princess Alice,
	Blooming Daisy,	Medora,	Princess Royal,
	Clarinda,	May Morn,	Park 2d,
	Derby,	May Belle,	Rossie,
	Heather Belle,	Miss Meikle,	
	Heather Bloom,	Mars,	
	Lady Avondale.		
Gibson's	Edith.		
Hungerford, Brodi & Co.'s	Ayrshire Lass,	Lady Ayr,	Red Rose.
Logan's	Greig.		
Morgan's	Adino,	Georgie,	Model of Perfect on,
	Annie,	Habbie Simpson,	Sea Bird.
Miller's	Daisy,	Favorite.	
Paterson's	Beauty,	Lord Raglan,	Geordie.
Peckham's	Highland Maid,	Sally,	Village Belle.
Peters'	Brenda,	Jean Armour,	Queen 2d,
	Duchess 2d,	King Coil,	Rosa,
	Harold (formerly Dr. Hornbook),	Merryton 3d,	Young Merryton 2d.
Sturtevant's	Domine,	Eos,	Mona,
	Drusilla,	Glengarry,	Ops,
	Edna,	Mains,	Selena,
Thompson's	Annie,	Duchess,	Rassie 2d,
	Crown Prince,	Highland Chief,	Rassie 3d,
	Diamond.		
Whitney's	Barrochan Maid,	Flora,	Nettie,
	Bessie Belle,	Jock,	Rassie,
	Bonnie Lassie.		
Walcott & Campbell's	Lady Clyde,	Rob Roy,	White Lily,
	Lady Glasgow.		

[1] This list includes animals imported in dam from prize-taking parents. Probably a large proportion of imported Ayrshires have taken prizes at some of the numerous Scotch fairs.

WOOD-CUTS OF IMPORTED ANIMALS.

ALTHOUGH the ordinary wood-cut of this animal usually conceals defective parts, and brings into undue prominence those forms which are deemed desirable, yet a study of these pictures brings to the mind an accurate idea of the shapes considered Ayrshire, by the artist at least, and a series of the same artist's pictures are accurate enough to be comparable with each other. The distribution of color is accurate, and this is something. These figures, then, if wooden in their look, and showing the animal in the best position, and under the most favorable circumstances, and if even exaggerated in parts, are of assistance to the breeder who brings to his work the preparation of study and reflection.

The following table of cuts of imported animals, although far from complete, yet may be of service.

Albert. Ag. of Mass. 1861, p. 21.

Ayr. Trans. N. Y. Ag. Soc. 1849, p. 84.

Baldy. Trans. N. Y. Ag. Soc. 1867, Part I, p. 35. Trans. Vt. Dairyman's Association, 1869-70.

Barrochan Maid. A. H. B. p. 89. Trans. Vt. Dairyman's Association, 1870-1.

Bonnie Lassie. A. H. B. p. 100.

Champion. Count. Gent., Sept. 12, 1872.

Cocksey. Count. Gent., Oct. 3, 1873.

Daisy. Count. Gent., Nov. 14, 1872.

Dandy. Trans. N. Y. Ag. Soc. 1849, p. 86.

Flora.   Count. Gent , June 11, 1874.
Geordie.   Farmer's Lib. ii, 385.
Georgie.   Ag. of Mass. 1873–4.   Ag. of Me. 1873.
Habbie Simpson.   A. H. B. p. 41.
Harold.   Dept. Ag. Rept. 1863, p. 194.
Handsome Nell.   Trans. N. Y. Ag. Soc. 1868, p. 182.
Jean Armour.   Ag. of Me. 1862, p. 61.
Jessie.   Ag. of Mass. 1861, p. 15.
Jock.   A. H. B., bet. Vols. II and III.
Lady-Ayr.   Trans. N. Y. Ag. Soc. 1860, p. 143.
Lady Kilbirnie.   Trans. Vt. Dairyman's Association, 1871-2.
Mars.   A. H. B. p. 56.
Medora.   A. H. B. p. 177.
Miss Miller.   Dept. Ag. Rept. 1863, p. 196.
Model of Perfection.   Trans. Vt. Dairyman's Association, 1871-2.
A. H. B. p. 181.
Netty.
Robbie Burns.   A. H. B. p. 67.
Rosie.
Rossie.   A. H. B. p. 204.
Tibbie.   Trans. Vt. Dairyman's Association, 1869–70.
Village Maid.   Count. Gent., June 19, 1873.

PEDIGREES OF IMPORTED ANIMALS.

(HELIOTYPE.)
PRIDE OF THE HILLS.

Calved May 7, 1871. Owned by John S. Holden, Belleville, Ont.

Imported in dam Barrochan Maid by N. S. Whitney, Montreal, P. Q., in September, 1870.

Barrochan Maid was bred by J. Holme, Japstone, Neilston, Scotland.

She gained the Silver Challenge Cup, valued at £25, at STIRLING, open to all Scotland, as the best Ayrshire, and the first prize at KILBRIDE, and Silver Medal for the best cow of all the prize cows.

(HELIOTYPE. PHOTOGRAPH OF A PHOTOGRAPH.)
LADY KILBIRNIE.

Owned by Sturtevant Bros., S. Framingham, Mass.

Bred by Robert Orr, Kilbirnie, Scotland; imported by James Brodie, Rural Hill, N. Y., in May, 1871.

GEORGIE.

Calved spring of 1866.

Owned by Sturtevant Bros., S. Framingham, Mass.

Bred by James Wilson, Boghall, Houston, Renfrewshire, Scotland.

Imported by J. N. Morgan, Ogdensburg, N. Y., in April, 1871.

Georgie, when a two-year-old, was first at HOUSTON; likewise gained medal for best cow in the yard, beating Barrochan Maid. When four, gained first prize at HOUSTON, likewise medal for best cow in the yard, and at RENFREWSHIRE COUNTY SHOW, she was again first, again beating Barrochan Maid.

MODEL OF PERFECTION.

Calved in spring of 1865.

Bred by Robert Wilson, Kilbarchan, Scotland.

Imported by J. H. Morgan, Ogdensburg, N. Y., in July, 1870.

Owned by Sturtevant Bros., South Framingham, Mass.

In 1869, after calving, she carried two first prizes at GLASGOW, amounting to £20, and a silver medal; likewise carried EAST KILBRIDE, first prize, and when nearly three months calved, was third at HIGHLAND SOCIETY. Before calving she was second at MAYHILL, first at BARRHEAD, and first at HAMILTON. The previous year, when three years old, she stood second at GLASGOW for cows of any age; at EAST KILBRIDE she stood first as a three-years-old in milk, etc. etc.

APPENDIX.

MILK:

Its Formation and Peculiarities,

WITH ESPECIAL REFERENCE TO

THE AYRSHIRE, JERSEY, AND DUTCH COW.

BY

E. LEWIS STURTEVANT, M. D.

AYRSHIRE, JERSEY, AND DUTCH MILKS:

THEIR FORMATION AND PECULIARITIES.

THE philosophy of breeding teaches that every observed effect must have been preceded by an adequate cause, and that intelligence and skilled observation may enable our reason to trace out the sequences which connect the one with the other with such exactitude as is permissible to our knowledge. It also teaches that inheritance is a form of force as uniform in its action, and as invariable, as is the force of gravity. Like gravity, its action is modified and interfered with by opposing forces, which disguise oftentimes its phenomena. As gravity acts alike on the feather and the bullet, so does inheritance act alike on all animals. In vitality we have such a complexity of phenomena, that a right interpretation is oftentimes difficult, if not impossible; yet the grand law of inheritance, the transmissal of qualities possessed by ancestors, may be disguised in individuals, but cannot be denied to the race.

It is to this universal law of inheritance, as modified by other laws, — the resultant of whose forces is the animal form, — that we are to seek the explanation of the variations that occur between members of the same species, breed, families, and individuals.

Those features of animal form that are readily cognizable, are usually more changed through the breeder's art, than other features which are not so readily noted. Consequently, the grazing breeds have been brought to a greater uniformity and perfection, than have the dairy breeds, as the changes to be desired have been more clearly indicated in the beginning, and recognized in the achievement. Changes in the dairy breeds are to be understandingly brought about by breeders and farmers, who have a practical belief in the universality of law,— that inheritance of form is not more important in modifying the shape of body than it is in determining the product from the animal.

Whether a cow's milk is better fitted for the making of butter or cheese, or for any other purpose, is largely determined by inheritance; as is also the amount she will give, the manner in which she will give it, the economy with which she will produce it from her food, and the effect of the production upon the health of the animal.

Milk is the product of the mammary gland, and is a fluid intended for the nourishment of the infant animal. It contains, therefore, all the elements needed for development and growth, and, chemically, is thus a perfect food.

The milk-glands, whose mammæ or teats furnish the name to the class *Mammalia* of naturalists, are four in number, in the cow, and, united by enveloping tissues, form the vessel called the udder. This organ occupies the posterior portion of the abdomen, bounded laterally by the thighs, and varies somewhat in shape, according to the breed or individual difference.

In the Ayrshire cow the glands of the udder are flattened, and held close to the body by a fibrous, and in part elastic tissue. The teats are small, cylindrical, and set wide apart. The teats are prolongation of the gland structure, in order to form an outlet for the secretion. As the gland is flattened, the affinity seemingly required by structure is, that the teat should be rather short and flattened, that is, cylindrical rather than cone-shaped. In the Jersey breed the glands of the udder are pointed and the teats are cone-shaped. They partake in form of the elongation of the gland. The glands are not held as close to the body as in the Ayrshire, but are pendent. The glands are seldom of equal size, the anterior ones often displaying a tendency towards extreme diminution, and the teats hang closely together. The American Holstein cattle — those large black and white cattle from Holland — have an elongated udder. There seems a hereditary want of tone in the tissues, as it is usually quite pendent. The glands are elongated, and in turn the teats are elongated cones.

The outer covering of the udder is composed of skin similar to that covering the body, but more thin and pliable, and is covered more or less with a fine hair of considerable length. Its interior structure comprises areolar tissue, and white fibrous and yellow fibrous tissue, which not only form septa between the glands, and connect the lobes, but also envelop the glands, holding them in position, and, by their elasticity and firmness, acting an important part in an organ subject to such violent changes of size. Fatty tissue occurs near the surface of the glands, and between their interstices, to a greater or less extent; and in the virgin heifer largely determines the form and size of the bag.

The teats, usually projecting slightly forward in the heifer, are likewise covered with a skin similar to that of the udder, but uncovered with hair, pliant, flexible, and creased. Their number corresponds to that of the glands, and they are interesting as forming the outlet for the secretion, as well as their mechanical adaptation to the needs of the calf. Their structure is an areolar and fibrous tissue beneath the skin, which, by its elasticity, closes the outlet and prevents the escape of the milk. Sebaceous glands are present, particularly at the base, and their secretion renders the surface soft, and less subject to injury.

The gland portion consists of ducts, reservoirs, glandules, and connective matter. The reservoirs are situated mostly at the periphery and apex of the gland, and more particularly adjacent to the sides covered by the skin of the animal. It is the reservoirs which cause the lobulated feel of the surface of the udder, in large part, and they serve to enclose the secreting surface, which principally occupies the centre. The glandules or vesicles, in their arrangement, form groups, and each group has its duct, which connects with the ducts from other groups, and thus the secretion is passed towards the main ducts, which serve to store and transmit what they receive. There is this peculiarity about the lactiferous ducts: they are not strictly uniform in size throughout, nor do they lessen or increase in size by regular gradation, but, slightly contracted at their inlet and outlet, have a bulge between, — thus, in form, a series of saccular cavities. The ducts and reservoirs are thus, in one sense, the same. The constricted portion of the reservoirs is formed of elastic tissue, which underlies the mucous membrane of the ducts. Thus, by the retardation of the milk, as it passes from the vesicles, where it is manufactured, towards the teat, its outlet, the pressure of the accumulated quantity is equalized to a certain extent throughout the gland. These ducts and pouches are lined by a vascular mucous membrane. On the exposed surface of the mucous membrane is a thin covering of tessellated epithelium to defend it from injury. As we reach the coecal extremity of the system we are describing, we observe the epithelium changing its character on the edge of the glandules or vesicles. The vesicle itself is lined with cells, which differ in size. These cells are the secreting portion of the gland, and by their own increase and casting off are themselves the morphological portion of the milk, the fat globule. These *acini* are surrounded by a net-work of capillaries, which form a *rete* or net on their surface, and furnish the blood for the use of these organs. Under the stimulus of abundant supply of force from this blood, these cells grow, and by a species of budding or proliferation, accompanied, it may be, by a species of fatty degeneration in their contents, the old cell is cast off to appear as the milk globule, while the new cell takes its place.

The milk globule is consequently formed from the animal; nay more, was, up to the moment of separation, a portion of the animal, subject to whatever changes may have been impressed upon it by its position, and formed through, and subject to, whatsoever changes may have affected it through its relation with the animal, as those arising from inheritance and environment. There is this difference, however, between these cells and the milk globule: In the one case, a portion of the animal, they are subject to changes impressed by the animal; in the other case, free from the animal, simply stored in the udder, they can receive none of these changes; they are as independent of their parent cells as when they are placed in the milk-pail.

According to Striker, fat globules may be detected in the *acini* of women who have died from puerperal fever. From careful observations on the *acini* from the gland of the cow's udder, we have been unable to detect separate fat globules in any one instance. We can say, however, with considerable confidence, that the cells from the *acini*, when detached, can in nowise be distinguished from the globule of milk from the same udder.

We will now allude to the uniformity of the plan observed in nature, the production of different results, rather by modifying parts already formed than by creating anew.

The mucous membranes may be considered as internal prolongations of the skin. The cells of the cuticle of the skin are colorless and flattened, often wrinkled and folded, and correspond to the pavement or tessellated epithelium of the mucous membrane. Subjacent to the epithelium or epidermis, there occurs a structureless basement membrane, which can rarely be demonstrated on account of its extreme tenuity. The third layer of the mucous membrane, corresponding to the *cutis vera* of the skin, is also composed of areolar and elastic tissues, and in both is highly vascular, and furnished with *papillæ* or *villi*. These three structures in both are supported by a layer of lax tissue, in which the *areolæ* frequently contain fat.

Glands themselves are of an epithelial nature, and are but adjuncts of the skin. As Virchow explains it, an epithelial cell begins to divide, and goes on dividing again and again, until by degrees a little process composed of cells grows inward, and spreading out laterally gives rise to the development of a gland, which thus straightway constitutes a body continuous with layers of cells originally external. Thus arise the glands of the surface of the body, the sudoriferous and sebaceous glands of the skin, and the mammary gland.

If, then, we could unravel the milk-glands, so as to present the interior surface flat, but little change except that of adaptation would be required to identify their structure with that of the skin. This is an important observation, as indicating the simplicity of method by which the purposes of nature are accomplished; and as a corollary to this simplicity, the effect of any agency, whether external or otherwise, on an animal, cannot be limited in its effect to one part only, but its influences must be more or less general in their nature.

We are now prepared to examine more particularly into the structure and reactions of milk, as we have seen that, through its method of formation, these must be influenced largely by the structure of the animal from which it is obtained.

Milk is one of the animal fluids which contains a morphological element, which in the form of myriads of minute globules of mixed fats, enclosed each in an enveloping substance, floats at will in a fluid composed of sugar of milk, caseine, etc., in solution. In this paper we shall consider only this globule, from which the milk derives its color and opacity, and which has sufficiency of form and character to be influenced by variation in breed and environment, and to influence itself in turn the character of those important dairy products, butter and cheese.

These globules are of varying size, some so small as to appear as granules under a magnifying power of 800 diameters, others occasionally attaining a size of 1-1500 of an inch. The small globules, for an increased power has invariably defined them as such, I shall for convenience term granules. As 1-27000 inch is, with my micrometer, a convenient division,

I shall speak of all globules less than this figure as granules, and all above as globules. Every sample of milk I have yet examined has shown these granules,[1] yet in some milks much more abundant than in others. In the skim-milk the granule has always been readily found, even in those milks where it was nearly absent in the cream. These globules being composed of various fats, surrounded by a pellicle, are intimately mixed with the milk as it comes from the cow; but their position soon becomes changed as they come under the influence of gravity, and they rise to the surface of the milk to form cream. As the weight to the covering of the fat globules, which is heavier than water, increases proportionately to the volume of fat as the sphere is diminished in diameter, the various globules show difference in physical action. When the weight of the covering is just or nearly sufficient to balance the low specific gravity of the fats, the globules remain nearly stationary in the fluid; when, however, the globule is large, the specific gravity of the mass is so much less than that of the fluid in which it occurs that it speedily reaches the surface. It therefore follows that the upper layer of the cream is composed of larger globules than the lower layer; or, giving expression to a general fact, the further you go from the surface of milk which has been at rest, the smaller the milk globule.

---

[1] That my conclusions may not seem to have been derived without study, I wish to say here that January 21, 1873, I found *recorded* in my note-book considerations involving the recognition of 9823 milk globules; and since that date many additional observations have been made. E. L. S., October 15, 1874.

Experiment I.

Three drops of milk were taken from a vessel containing milk which had been undisturbed for fourteen hours:[2] —

1st drop. Top layer cream............Average size of globule, 6120″.
2d "   Lower layer cream..........Average size of globule, 6640″.
3d "   Six inches below surface....Average size of globule, 8260″.

Should the globules which occur at these different depths be churned, it would be found that the different layers would require a greater or less exposure to the churning action to produce butter, and the butter would vary somewhat in quality in each churning. This may be readily verified by skimming a vessel of milk at intervals, and churning the cream of each skimming by itself.

The process of churning consists in breaking the covering of the milk globule and collecting the released fat into lumps. This breakage seems usually to occur through friction; and the ease with which it occurs is determined in part by the toughness of the investment, and in part by the size of the globules.

In general, the time required for churning milk or cream from the same breed, into butter, has a close relation to the size of the globule.

Experiment II.

Three Jersey cows, on similar feed, yielding same amount of milk. The milk of the same milking, set

---

The sign ″ signifies ths of an inch To illustrate: these figures are to be read 1-6120th of an inch, 1-6640th of an inch, etc.

on the same shelf, and the cream churned as nearly as possible at the same sime, by stirring in a pitcher with a spoon: —

Name of Cow.	Average size of globule.	Time of churning.
Desdemona	4440″	13 minutes.
Gazelle	5260″	30 "
Beatrice	5520″	34 "

Experiment III.

The milk, except when otherwise stated, was in this experiment fresh from the cow, and cooled to 60° by immersing the Florence flask, used as a churn, in cold water: —

Average size of globule.	Time churned.	Butter first showed.
5680″ (cream)	3 minutes	—
5940″	8 "	5 minutes.
6768″	25 "	15 "
8252″ (churned with egg-beater)	50 "	36 "
8320′	60 "	—

Having established the fact that the size of the globules determine some of the reactions in the churn, we will consider the effect of churning milk containing globules of widely different sizes. Whenever such trials have been made and the results carefully noted, I have found that the larger globules become divested of their covering first; and oftentimes, I suspect, being overchurned, hinder the same process going on with the same facility for the rupturing of the smaller globules. The overchurning of butter destroys the grain, or the natural form in which the butter is contained in its investing covering, and pressing out the oleine, as I conjecture, furnishes to the fluid this oil in emulsion, which decreases the friction to which the globules are subjected in the process of separating butter. The butter product is thus, theoretically at least, diminished, and its churning retarded.

Experiment IV.

Carefully measured 16 fluid ounces of milk fresh from the cow and cooled to 60°. After twenty minutes' churning, the butter was collected by straining the fluid through fine linen. The amount, 57 grains, or a proportion of one pound of butter to about 60 quarts of milk. The next day churned the buttermilk. After an hour and a quarter's agitation, 211 grains of butter were collected.

This milk threw up twelve per cent of cream, and was therefore of good average quality, as was also indicated by the butter proportion of one pound of butter to about thirteen quarts of milk.

We must seek an explanation of this experiment in the physical reaction of the globules.

Average of ten measurements of the globules occurring in a line 1-100 inch in length:—

Top layer of Cream.	Lower layer of Cream.
6345″	.8180″
6300″	.6390″
6255″	.8505″
6480″	.8100″
6010″	.7155″

The granules in the lower layer were very numerous, but not considered in forming our averages.

The impression gained on observing this milk microscopically was a great variation in sizes of globules, so much so as to suggest a division into two classes, as if two different globuled milks had been mixed.

If the measurement of 100 globules of the cream may be taken as giving an indication of an average, we had 24 globules larger than 6750″, and 76 globules of that size and smaller, — a proportion of about 1 to 3. The proportion of butter in the results of the two churnings was about 1 to $3\frac{1}{2}$, a correspondence sufficiently close to be suggestive, and, taken in consideration along with the microscopic investigation of the butter-milk, offers the explanation that the larger globules principally furnished the butter of the first churning, while the smaller globules were the principal factors in producing the butter in the second churning.

The experiment can be verified in a very simple way by shaking some milk in a clean white glass bottle. After a short time specks of butter will be seen adhering to the glass, the product of the breaking of the large globules, while it may be a long time before the butter will appear in the ordinary acceptance of practice.

Another consideration in the study of the globule is the effect of the distance of the cow from calving, on the size. As a constant result with me, the further from calving the smaller the globule, and I think the more uniform the sizes.

Experiment V.

The milk of the same cow at various periods from calving: —

Days from calving....1½ Average size of globule....4400"
"   "   " ....3½   "   "   "   ....4666"
"   "   " 33   "   "   "   ....6000"

Experiment VI.

Three cows of the same herd, and under the same treatment. The trial was made with milk of the same milking, treated alike:—

No. 1. Days from calving, 15. Average size of globules, 4440"
" 2.   "   "   "   27.   "   "   "   5260"
" 3.   "   "   "   40.   "   "   "   5520"

No. 3 had a great uniformity of globule, and very few granules. Except for the granules, No. 1 had been as uniform. That the "feed" did not probably affect the experiment unfavorably in this case, I give below not only the food but the proportion of butter to milk.

Butter to Milk.
No. 1. Pasture and ½ qt. oil meal, about 2 or 3 qts. shorts, 1 lb. 23.23 lbs.
" 2. Pasture and 3 qts. oil meal, 3 qts. shorts, 1 qt. oats, 1 " 23.27 "
" 3. Pasture and ½ qt. oil meal, about 2 to 3 qts. shorts, 1 " 17.77 "

Experiment VII.

Milk of different cows, but of the same breed. Measurements taken at different times, and under varying conditions of food, etc. The sequences are not therefore as regular as in the Experiments V and VI.

Days from calving, 1½.......................... 4935"
"   "   "   3½.......................... 4718'
"   "   "   12 .... 5580"
"   "   "   33 .... 6384"
"   "   "   60 .... 5400"
"   "   "   135 .... 6040"
"   "   "   375 .... 6339"

By including some measurements which were taken from the lower layer of cream, and not incorporated in the above table, we have additional illustration.

Days from calving,	1¼	4580″
" "	33	6200″
" "	69	6750″
" "	135	6720″
" "	375	7660″

The size and appearance of these globules is varied, as I believe, by the feed of the cow, and certainly, to a considerable extent, by her condition. That their size has a connection with the grain of the butter, it is in the power of any one to convince himself by direct experiment. The larger-globuled breeds furnish butter of a stronger grain than do the smaller-globuled breeds, and the first rising from the milk-pan yields also a stronger-grained butter than does the succeeding risings.

When a cow gets out of condition she oftentimes falls away in her milk very rapidly, and a microscopic examination of her milk may show the presence of colostrum corpuscles. In order to understand the signification of this fact, it is necessary to know what colostrum is.

Writers upon milk have made statements of wide discrepancy. Dr. Bird[3] states that the colostrum of the cow is yellow, mucilaginous, and occasionally mixed with blood; it contains but mere traces of butter or other fat, and appears to contain albumen as one of its ingredients. This secretion does not

---

[3] Cooper, Anat. of the Breast, p. 124.

turn sour like milk, but readily putrefies. According to Stiptrian, Luiscius, and Boudt,[4] however, the colostrum from the cow yields 11.7 per cent of cream, 3 of butter, and 18.75 of cheese. Thomson[5] states that colostrum when churned gives a very yellow butter, which, when heated, emits a smell similar to the white of an egg. Heine and Chevalier[6] give 15.1 per cent of casein, 2.6 of butter, and 2.0 of mucus. According to Lehman,[7] the colostrum is richer in fat than the corresponding milk. In the analysis by Boussingault,[8] mention is made of 3.6 per cent of sugar of milk, — a substance entirely unmentioned by Heine and Chevalier, and the other authorities we have quoted above.

According to Beale,[9] colostrum contains many large cells, consisting of an investing membrane filled with oil globules resembling those which are floating free in the surrounding fluid. Donne[10] states the colostrum corpuscle to be made up of small granules, united together or enclosed in a transparent envelop. He says they disappear in ether, and that he traced these globules in milk secreted twenty days after parturition. M. Guterbock[10] has also observed these compound globules, and says he could detect the transparent membrane after the ether had dissolved the enclosed granules. M. Mandl[10] has not been able to detect these compound globules, and believes them to be made up of agglomerated milk-globules.

---

[4] Cyc. Anat. and Phys. iii, 360.
[5] An. Chem. p. 435.
[6] Johnston's Chem. p. 535.
[7] Phys. Chem. ii, 64.
[8] Journ. R. A. S. of Eng. xxiv, 301.
[9] The Microscope in Medicine, p. 267.
[10] Cyc. Anat. and Phys. iii, 361.

Kolliker[11] thinks the formation of colostrum the introduction to that of milk. He also thinks that the colostrum may be the product of a degeneration, and thinks that is in part derived from the internal cells of the originally solid rudiments, which are removed. Virchow[12] states that it is the still coherent globule, which results from the fatty degeneration of an epithelial cell. According to Reinhardt[13] they are transformations of the epithelial cells of the mammary ducts, the result of a sort of fatty degeneration or regressive metamorphosis, consequent upon the peculiar activity of the mammary gland during pregnancy.

When we consider the physiological formation of the milk-globule, as set forth in this paper and elsewhere[14] by the writer, the relations of the colostrum corpuscle to the milk will be readily noted. In the earlier stages of lactation, and before parturition, the process of the casting forth of these milk-cells is not perhaps so complete as at a later stage. We have at first a tardiness or lack of co-ordination of action between the different cells; action stimulated by the uterine function is with cells which have been for a long time stationary or in partial rest. Therefore, through an imperfection of process, groups of cells from the same vesicle are forced off in mass before they are ready to become milk-globules. When lactation commences the whole structure of the udder

---

[11] Human Histology, li. 279.
[12] Cellular Pathology, p. 376.
[13] Carpenter's Human Phys. p. 818.
[14] See the forthcoming Prize Essay of the New York State Ag. Soc. of 1873, "Milk." Also, Ag. of Mass. 1873–4, p. 376.

glands is in a state of wondrous activity. We have such an excess of action that cells are cast off from the vesicles prematurely, and tear off the adjoining cells while contact still exist. The colostrum corpuscle is theoretically but a portion of the lining of the milk-vesicle, detached before the cells have arrived at the stage of the milk-globule. It is but a stage of development of milk-globules. It may be caused by excess or defect in nutrition, through any cause which may produce an imperfect development of the cells of the *acini*. Hence, when sickness overtakes a cow, even at a long period from calving, colostrum corpuscles may appear in her milk.

By careful microscopic observation, I have been able to detect no difference between the globules present in the colostrum and the milk-globule present in the milk. The action of ether, in my hands, does not cause the globules to disappear, although it may have some action on the granules. I sometimes have noticed a change produced, without my being able to define exactly what the change was.

The specific gravity of cream must be subject to considerable variation, according as the globules vary in size and the thickness of the investing coating. As writers have experimented with milk from different breeds of cows, and under different circumstances, it is no wonder that their results are discordant, but it is unfortunate that we do not have sufficient particulars to enable us to place the reason in the right place.

The specific gravity of cow's milk is said to be lighter than milk but denser than water, by Dr. Voelcker,[15] who gives the following as the result of his trial, and Willard[16] accepts these results.

```
From milk after standing 15 hours........ ...1019.4 at 62°
    "        "         "   43   "   ............1012.7  " 62°
    "        "         "   48   "   ............1012.9  " 62°
```

Letheby, in his Lectures on Food,[17] states the specific gravity at 1013, while Berzelius[18] places the specific gravity of cream at 1024.4, the figures which are accepted by Dr. Golding Bird.[19] L. B. Arnold,[20] of Rochester, N. Y., states that cream has the specific gravity of 985.

In my own experiments I have usually found that a drop of cream, carefully dropped on rain-water, would float. It even floats when dropped into the water from a height, so that the force of the impact carries the drop below the surface or spreads it on the surface. In one instance only have I known the cream to sink when carefully placed on the surface of rain-water.

Experiment VIII.

In one carefully-conducted experiment made with the cream from the surface of a large cream-jar, I found the specific gravity to be 983 at 62° by weighing.

---

[15] Journ. R. A. S. of Eng. 1863, pp. 298, 317.
[16] Dairy Husbandry, p. 168.
[17] Ibid. p. 34.
[18] Johnson's Farmers' Enc. pp. 240, 814.
[19] Cooper's Anat. of the Breast, p 119.
[20] Am. Dairyman's Ass. Trans. 1870, p. 160.

Experiment IX.

A microscopic investigation would seem to settle the question that some cream may be lighter, other cream heavier than water. I added some milk to water in a tall glass tube. The milk all fell rapidly to the bottom, in a smoky cloud, leaving the upper third of the tube absolutely clear. In half an hour the density at the bottom had diffused itself upwards, in a regular gradation of opacity, even to the top. Upon examining a drop from the surface of the water, under the microscope, globules showed of quite even sizes, ranging generally from 6750" to 4500" in diameter. A like examination of the bottom layer scarcely showed a globule larger than 9000" in diameter, yet two globules were seen as large as 6750".

Having now considered the formation of milk in respect to one of its constituents, and treated the subject in a general manner, we are now prepared to examine into the peculiarities which come of breed and are consequent thereto.

My opportunities have been limited to three breeds, — the Ayrshire, the Jersey, and the Dutch, those large black and white cattle from Holland, the American "Holsteins." Although such differences as are to be discussed are probably of universal application, yet here my conclusions will be confined to the result of my own examinations, which have been fairly complete with reference to the Ayrshire and Jersey milks, but more limited with the Dutch.

The breed of the Ayrshire cow furnishes a globule intermediate in size between the Jersey and the Dutch. The prominent feature of this milk is the numerous granules.

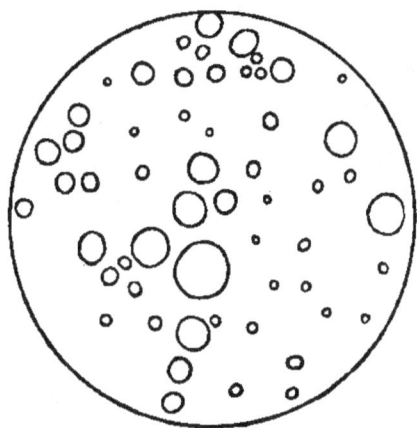

×813

Upon a careful examination of Ayrshire milk, we find an indication of a division of the breed into two classes, according as they have been bred for butter or cheese purposes. When we examine the milk which characterizes the type of the two classes, the differences are manifest and the peculiarities are readily noted; but these two types shade into each other so imperceptibly and gradually — like two separate, overlapping clouds — that the division line is obscured. Those cows which appear the nearest to such a line may be grouped as a third class.

The typical butter family of Ayrshires furnishes milk possessing a globule scarcely inferior to the Jersey globule in size, yet the sizes are more varied, and granules present in abundance. The skim-milk is not as blue as Jersey skim-milk on account of the presence of the granule. The envelop to the globule seems tougher than in the Jersey globule, and the milk takes a somewhat longer time to churn. The effect of the acids developed in the milk by keeping also appears to affect the churning qualities of the milk to a less degree. Therefore, although the Jersey milk may be skimmed, certainly not later than when the milk commences to thicken or "lobber" at the bottom of the pan, the Ayrshire milk should pass considerably beyond this point, and develop somewhat more acidity, before the cream is removed.

Experiment X.

Average size of globules.	Cream.	Churning.	Butter came.
Ayrshire........4666"	3½ days old.	20 minutes.	—
Jersey..........5680"	Old.	3 "	—
Ayrshire........6000"	New milk.	25 "	15 minutes.
Jersey..........5940"	" "	8 "	5 "

In the new-milk churning, when the globules of the two breeds were about of a size, the Jersey milk churned much the quicker, probably on account of the thinness of the enveloping membrane of the globules. In the Jersey cream we also have a similar result, although the globule is smaller than in the Ayrshire cream, with which it is compared.

The butter from the Ayrshire cow is of good texture; is yellow, often a deep yellow, but, as far as I have observed, not possessing the peculiar orange tinge of the Jersey.

The typical cheese family of Ayrshires furnishes a milk of much smaller globules and more numerous granules than the butter type of Ayrshire milk. The milk throws up a small percentage of cream, and is specially fitted for the manufacture of cheese, as the theoretical essential for the best result in cheese-making is, that the butter should be retained in, and evenly distributed through the cheese. When cream rises, in the ordinary process of manufacture, it does not again readily mix with the milk, but much of it passes off in the whey. When, therefore, the milk is rich to analysis, but the cream percentage is small, on account of the butter-globules being too minute to rise very rapidly, or at all, through the fluid, then we have milk conditioned for the most favorable results. I do not question but that, by the means of the microscope, milk could be selected which would endure reasonable skimming, or that amount of skimming which could take place in ordinary cheese-making, and yet make a richer cheese than another selected milk, which might contain fully as much fat, and be used unskimmed.

In order that this statement may be rendered clearer, let us see upon what conditions, in part, the character of cheese depends. It must be borne in mind that, if these observations of mine are correct, as they surely are, the dairyman deals not alone with composition of milk, but also with structure, in the processes of either butter or cheese making.

During the ripening of cheese a portion of the caseine or curd suffers decomposition, and is partially changed into ammonia; the latter, however, does not escape, but combines with fatty acids produced in course of time from the butter. The peculiar mellow appearance of good cheese, though due to some extent to the butter which it contains, depends in a higher degree upon a gradual transformation which the caseine or curd undergoes in ripening.

Such being the process, it is quite evident that an even distribution of the fatty matter through the curd, is desirable, in order that each particle of ammonia, as set free, may at the moment be in contact with the fatty acid which is supplied from the fat globule. Consequently that milk which contains the cream in a state of equilibrium throughout the fluid, and yet which is rich by analysis, fulfils best the desired conditions.

That the facts of dairying are in accordance with these views, witness a few statements. Dr. Voelcker writes that one of the chief tests of the skill of the dairy-maid, is the production of a rich tasting and looking, fine-flavored, mellow cheese from milk not particularly rich in cream. That this can be done is abundantly proved by the practice of good makers. In the accounts of cheese-making that come to us through the Transactions of the Cheese-Makers' Associations, we find both concordant and conflicting testimony, which can only be rendered concordant by the supposition that the parties reporting, of equal repute, used milk of different characters. Thus some makers advocate taking the cream of one milking for the purpose of butter-making, and deny any injurious influence therefrom on the cheese, while others deprecate this course. Many others think the cream may be profitably removed in the fall, but not at other times. When we consider that the cows which furnish the milk to a factory usually calve in the spring, and that the milk-globule diminishes in diameter with the time from calving (see Experiments VI and VII), the reason underlying this cause may be seen to reside in the character of the milk differing with the season.

Mr. Gardner B. Weeks has sold from his creamery skim-milk cheeses in quantity at a price within a cent and a half a pound, of the highest quotations of whole-milk cheeses. All writers unite in testifying to a loss of butter in the whey, and processes are patented for the extraction of this waste butter for family use. These considerations concerning the milk-globule, point out the way to prevent waste, and to obtain full price, by regulating the character of the milk supplied, or manufacturing in accordance with the character of milk supplied, rather than other more wasteful alternatives.

The milk of the Ayrshire cow, which holds a middle position between these extremes, is well fitted by its structure for either butter or cheese without being equal to the animal of the typal extremes for either product alone. The figure given illustrates the milk of cow of this third class, as I have called it. This class of milk is perhaps the most predominant, and is perhaps of the most value for the majority of farmers. It furnishes an excellent percentage of cream, — from 12 to 19 per cent, in our experience, — a good quality of butter, and a skim-milk of excellent quality. The skim-milk is neither as blue as in the butter type nor as white as in the cheese type of cow, but occupies a medium position,— the practical differences between these three types of milk being the greater uniformity of constitution of the milk, after standing, in one case than the other; the difference in the rapidity and completeness of separation of the butter-globule, or cream; the greater or less occurrence of the granules, or extremely small globules.

The milk-globule of the Jersey breed is larger than is the corresponding globule of the other breeds here considered, and there are fewer granules.

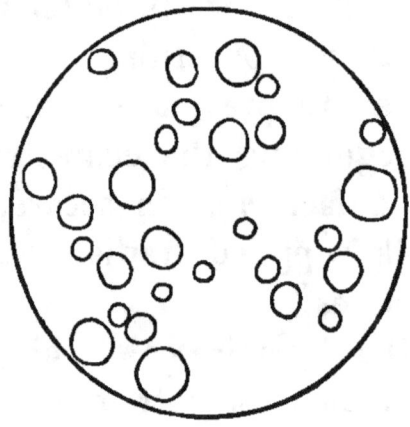

×813

The envelope to this globule seems weaker than the corresponding envelope in the other breeds, and more readily ruptured in the churn. (See Experiment X.) It is also more readily acted upon by the chemical changes induced in the milk by time. When the old cream of these breeds is examined microscopically, it is found that the Jersey globule is more readily broken or distorted by pressure than the others. Practically, therefore, this milk should be skimmed at an earlier period of the souring change than should the other milks. I feel assured, from impressions gained from my own experiments, that the Jersey milk should be skimmed certainly not later than when the milk commences to thicken or "lobber" at the bottom of the pan, while the other milks may pass considerably beyond this point with advantage.

From the large size of the Jersey globule, and the comparatively small number of granules, the Jersey cream rises with considerable rapidity, and so completely as to leave a very blue skim-milk. I have known the whole of the cream, in one sample of Jersey milk, to rise to the surface in four hours, but such rapidity is exceptional.

As the variations between the time occupied in churning, are determined largely by the milk-globule, we find that the cream with the largest globule takes less time to churn, than does a small-globuled cream. The size of the globule also determines the grain of the butter, while the breed determines to a large extent the composition. We hence find in the Jersey milk an aptitude to churn very quickly, under favorable conditions, and the butter produced to be of a waxy and strong-grained appearance. The butter is usually, perhaps always, colored by an orange pigment, which seems characteristic to the breed. Owing to this orange tinge of the pats, and the character of the substance investing the globule, the Jersey cream oftentimes appears yellow, especially after standing. This color to the cream is not peculiar to the Jersey breed, but seems more usually present, or more prominent in this breed than in the others.

When Jersey butter is shaken with boiling water, and the nitrogenous matter enclosed washed out and collected, it is found to be much more abundant than in Ayrshire butter, and of a somewhat more flocculent character. Hence, theoretically at least, Jersey butter should not possess "keeping quality" to such an extent as the other butters. (See Experiment XIV.)

The conclusions to be gleaned in reference to the Jersey milk are: First, that it is unfitted for the retail dealer on account of the rapidity with which the cream rises, and the difficulty of again mixing this cream with the milk (see Experiment XIII), and, on account of the absence of granules, the inferior quality of the skim-milk. Second, that on account of the completeness of the separation of the cream, it is an excellent milk for the butter-maker, exhibiting but little waste, and, with quick churning capacity, supplies a butter of excellent appearance and quality. Third, that on account of the physical qualities described, it is not an economical milk for the cheese-maker. Fourth, from the presence of nitrogenous matter in intimate mixture with the butter, the indications are that this butter is better fitted for the daily sending to market, than for the purpose of winter packing.

As an interesting observation, I would say that from the following and other experiments I have come to the conclusion that a judgment can be formed of the depth of color the cow will give to her butter by the examination of the wax secretion in her ear.

There is a striking resemblance between the various glands in the plan of their formation; and here, if anywhere, we should, *a priori*, expect to find correlations. It must be remembered, however, that the wax in the cow's ear changes color by exposure to the air, and consequently a freshly-exposed surface must be examined in the use of this indication. So great is the resemblance between the ear-gland and the milk gland, that in one case at least I have found a similarity in the size of the fat-globule in either.

Experiment XI.

No. 1. Very yellow in skin and ears.................... Guernsey..2d highest colored butter.
No. 2. Skin of udder not as yellow as No. 1. Ears as yellow...................... Jersey....Highest colored butter.
No. 3. Skin middling yellow, do. ears .... " ....Good color to butter.
No. 4. Skin rather light colored....................... " ....Rather light colored butter.
No. 5. Skin and ears yellow.. Ayrshire..Yellow butter.
No. 6. Skin and ears scarcely showing color.............. " ..White butter.

Nos. 1, 2, 3, and 4 were on similar feed, and the experimental butter was made from the same milking, at the same time. Nos. 5 and 6 were on similar feed, and the cows were selected on account of the variation in the color of the skin. The effect of the character of the food on the color of the secretions was well illustrated by cow No. 6, which usually has possessed more color of skin than at this trial.

The melting point of Jersey butter, as obtained by

me, has varied from 93° to 98°, from different herds at the same season of the year.

Experiment XII.

No. 1 herd	98°
No. 2 "	96°
No. 3 "	94°
No. 4 "	93°

In order to obtain the melting point of butter, the best process that I have yet found, is by the use of mercury. Heat a small plant-pot of sand to about 120°, and set in the sand a small cup of mercury, with the bulb of a thermometer immersed therein, and supported by a cross-bar. Having previously filled a section of a quill or a cylinder of paper, open at both ends, with butter, impale on a needle so that the point of the needle shall extend through a quarter of an inch. By immersing the quill or cylinder in the mercury, the projecting needle keeps the apparatus always at the same distance below the surface of the mercury, and the butter enclosed in the cylinder is subject to a uniform pressure of say three eighths of an inch of mercury. The moment the melting point of the butter is reached, the warm mercury forces it out; it immediately jumps to the surface, and at the same instant the observer reads off the height of the immersed thermometer.

Of the three breeds we are considering, the American Holstein presents the smallest globule to its milk. The globules are more uniform in their size than in the Ayrshire milk, and there are fewer granules.

The cream, on account of the uniformity of size of the globule, rises completely, and on account of their small size mixes again with the skim-milk with considerable readiness. The absence of granules as a pre-

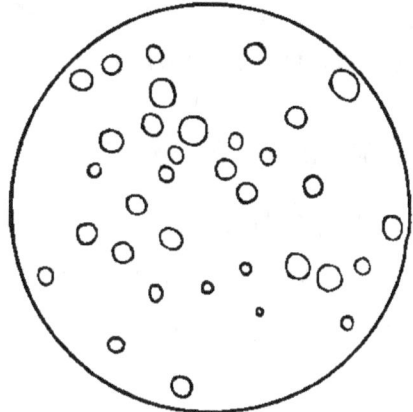

×813

dominant feature makes the skim-milk appear blue, and renders this milk less fitted for the cheese-maker than Ayrshire milk. The quality it possesses of the cream and skim-milk being readily miscible may offset in some degree the absence of the granules.

The butter made from this milk, so far as determined by a single experiment, was fine in grain, light in color, and displayed remarkable keeping quality. Perhaps the "keeping" power is the direction of the usefulness of this breed. My experiments with the milk of this cow have, however, been of too limited a nature to allow me to dwell very particularly on my results.

We will now compare the milk of the three breeds, and summarize in part our preceding showings.

Experiment XIII.

Milk from each of the three breeds was placed in bottles and the cream allowed to rise, the bottles being kept corked, to prevent evaporation from taking place to an extent sufficient to harden the surface of the cream. By shaking the bottle it was found that the Dutch cream mixed again with the milk with the greatest facility; the Ayrshire cream, less readily; the Jersey cream, with difficulty and imperfectly.

Experiment XIV.

One sample of Dutch butter, one of Guernsey butter, seven of Jersey butters, and three of Ayrshire butters, were placed in a cupboard adjoining a steam-heater. A few days later another pat of Ayrshire butter was added.

The Guernsey butter was very high-colored, melting point 99°, had an oily rather than a waxy look, but was very attractive. It moulded in spots in about a month.

In seven weeks the Jersey butters were all rancid, and one had lost its color in spots, the white spots reminding of tallow, — no butter flavor.

The Ayrshire butters were not rancid, but had lost flavor and were poor. The last specimen placed in the same cupboard, but on an upper shelf, was forgotten. When examined three and a half months later, it still retained its butter flavor and taste, but was not strictly first-class.

The Dutch butter was well preserved, being neither rancid nor flavorless.

The butter from cows of the *same* breed and on *similar* feed, and giving the *same* quantity of milk, made from the *same* milking and at the *same* time, does not necessarily present the *same* color. The color for the Jersey breed, I think, is yellow, more or less deep, and tinted with orange. That from the Ayrshire cow is yellow, often a deep yellow, yet, so far as I have observed, the orange tinge is lacking. The Dutch butter, speaking from several samples only, is light yellow, or a darker yellow, of attractive clearness.

The predominant feature of the Ayrshire milk, from whatsoever class it may be taken, is the presence of numerous granules or extremely small globules, which give a white rather than a blue appearance to the skim-milk.

The predominant feature of the Jersey milk is the size of the globules, the tenderness of their investing membrane, and the small quantity of granules. The skim-milk is hence blue, and does not readily remix with the cream upon agitation.

The Dutch milk has for a predominant feature the uniform yet small size of the globules, and the comparative absence of the granule. The skim-milk is blue, yet the cream can be readily mixed with it by shaking.

A curious feature brought out by experiment is, that the mixed milk from two breeds did not produce as much butter as would the same milk churned separately. When a large-globuled milk and a small-globuled milk are churned together, the larger globules separate first into butter, and the breaking of the smaller globules appears to be retarded. Moreover, the covering to the globules being of different character, those of one breed are ruptured more readily than those of the other, and over-churning of a portion of the product is inevitable. Think of churning Jersey milk, which will make butter in eight minutes (see Experiment X), mixed with Dutch milk which requires an agitation during sixty minutes (see Experiment III) for the same produce.

Experiment XV.

Two samples of milk were selected which showed considerable variation in the size of the globules. Twenty fluid ounces of the Jersey milk were divided into two parts; the like quantity of Ayrshire milk was similarly treated.

Jersey milk.   Average size of globules........ 5852"
Ayrshire milk.   "   "   "   .... 7080"

These milks were then cooled to 60°, and churned by shaking in a Florence flask.

Ten ounces Jersey milk. Butter came in five minutes; churned eighteen minutes. Product, one hundred and thirty-six grains of butter.

Ten ounces of Ayrshire milk. Butter came in twenty minutes; churned thirty minutes. Product, 76 grains of butter.

Thus the 20 ounces of milk churned separately produced 212 grains of butter, or a proportion of one pound of butter to 44.75 pounds of milk.

Ten ounces of Jersey milk plus ten ounces of the Ayrshire milk mixed and churned in like manner.

Butter came in thirteen minutes; churned twenty minutes. Product, 179 grains of butter. After this butter was removed the buttermilk was churned ten minutes longer, without producing any change in the result. The proportion in the mixed milks is, therefore, one pound of butter to 48.88 pounds of milk.

Difference in favor of churning each milk separately, 33 grains, or 4.13 pounds in the proportion.

When, therefore, a Jersey cow is kept in an Ayrshire or Dutch herd for the purpose of influencing the color of the butter, it is probable, in churning the produce of the herd, that the large globules of the Jersey milk are broken first in the churn; and while the smaller globules are being broken, the butter which first came is being over-churned, and theoretically at least the quality of the result is impaired, if not the quantity lessened.

When a few Ayrshire or Dutch cows are kept among a herd of Jerseys, and the milk churned together, we should expect, both theoretically and practically, a large portion of the butter of the small-globuled milks to be left in the buttermilk in the form of globules.

A like application may be made to herds of native or grade cows. Unless there be uniformity within certain limits, in the milk-globule, there is a loss of product. When uniformity is so seldom found in external shapes, as in a herd of natives, it is not probable that any greater uniformity exists between their functional productions.

The bearing of these facts of the physical construction of the milks, on practical questions, such as the depth of setting milk for cream, etc., are obvious. It is unsafe to arrive at empirical conclusions, and enunciate such as a law, when scientific conclusions, which give the reasons, are to be attained. Thus in reference to deep cans for the butter dairy: with Jersey milk, when the cream rises rapidly, they *may* be the best; but with other milks coagulation *may* occur before the smaller globules have reached the surface. Again, the quality of the cream of the different risings is widely different in the churn. The one method may furnish more cream, yet no more butter than the other.

It is thus seen how both those who claim and those who deny the benefits of deep setting of milk may be equally right from the standpoint of their own practice, while both may be equally wrong in applying their conclusions to other people's practices, for the results are largely brought about by the physical conformation of the milk, — a sufficient cause for differing conclusions, and a cause whose influence has thus far been entirely overlooked, in dealing with such apparently simple, yet really complex problems as arise in dairy practice.

As the milk-globule is determined as to size and quality in great part by inheritance, it is thus seen that there is a close connection between the breeders' effort to improve stock and the manufacturers' effort to improve the make of his cheese or butter. Minute differences often produce appreciable results; and he

Finale. — Milk Requirements.

For Butter. — That the globule should be of good size, of uniform size, and should be in abundance; *i. e.* a large percentage of cream.

Requirement best fulfiled by the *Jersey, Ayrshire,* and *Dutch,* in the order given.

For Cheese. — That the globule should be so small as to remain mixed with the milk under all circumstances; *i. e.* a white and not a blue skim-milk.

Requirement best fulfilled by the *Ayrshire.*

That the globules should easily remix with the milk after separation.

Requirement best fulfilled by the *Dutch* and the *Ayrshire.*

For the Milk Retailer. — That the globule shall remain for a sufficient period mixed with the milk, so that an evenness of quality may occur during delivery to customers.

Requirement best fulfilled by the *Ayrshire* and *Dutch.*

Farmer's Requirement. — An abundance of yield under given circumstances.[21]

Requirement fulfilled in the order *Ayrshire, Dutch,* and *Jersey.*

---

[21] Note that this application is that which is shown under the circumstances of same locality and known treatment.

CREAM.

When milk as drawn from the cow is allowed to stand, there is immediately a change in the relative position of the milk-globules in the fluid. These globules, so fine as to be in a state of equilibrium in the fluid, or whose specific gravity differs so slightly from the fluid in which they are suspended that their position remains practically unchanged, retain their distribution, and, after a time, become a constituent of the skim-milk. The remainder of the globules seek the surface of the milk to form cream, with a rapidity proportional to their separate specific gravities.

On account of the differences in the size and specific gravity of these globules, there is a tendency towards an arrangement of the cream in layers, the largest spheroids being at or near the surface, the smallest against the under side of the cream. In consequence of this arrangement, we find in cream an uneven product, as it is formed on the milk, each layer presenting a different-sized globule, and conse-

---

A paper read at the American Dairymen's Convention, Utica, N. Y., January 12-14, 1875, by E. Lewis Sturtevant, Waushakum Farm, South Framingham, Mass.

quently, as I have elsewhere shown in my writings, presenting a different reaction in the churn.

Now, each layer of the cream being different and producing a different character of butter, it is evident that one layer must be better for butter-making than another. It has been so determined by a series of microscopic and practical experiments combined, through which it may be stated as a rule, that the larger the milk-globule the quicker the churning and the better the butter, other things being equal. Hence, in practice, the first cream that rises on any milk is the richest; that is, it produces the best butter, and this butter churns the quickest. The second skimming furnishes cream poorer for manufacture, and the last skimmings may be worthless for high-class butter. Hence, in practice, a dairyman may obtain too much butter from his milk, the increase in quantity not sufficiently compensating for the decrease in quality, brought about through the churning of globules which should have been left in the buttermilk.

We recognize a liability in any butter to vary in manufacture from week to week, or possibly from churning to churning. There is often great faith pinned to special churns and to special modes of practice. Did it ever occur, that cream is a complex substance, scarcely alike in any two specimens, and is affected not only by the circumstances affecting its rising, but also by the food and condition of the cow?

SPECIFIC GRAVITY OF CREAM.

What is the specific weight of cream?

Berzelius,[1] an established authority on chemistry, says, 1024.4. Dr. Voelcker[2] says, 1012, 1019. Letheby[3] 1013, and Dr. Hanneberg,[4] of Stockholm, 1004.9 and 1005.5. It is an American authority, Prof. L. B. Arnold,[5] who obtains a result as low as 985.

In my own experiments, using cream from the top of a cream jar, I have obtained a specific gravity of 983 by weight; and on the other hand, I have found cream which would sink in water. It must seem exceptional and strange that such an apparently simple product as cream should show such wide discrepancies. If authorities are right, then cream is not as simple a substance as it is ordinarily described.

We will now examine some analyses of cream, and observe what the results teach us.

ANALYSES OF CREAM.

	Water.	Solids.	Butter.	Caseine, etc.	Sugar.	Ash.	Etc.
Mixed Cream[6]	59.25	40.75	35.00	2.20	3.05	.50	
Country Cream[7]	49.00	51.00	42.00	4.20	3.80	.60	
Jersey Cream[7]	36.40	63.60	56.80	3.80	2.80	.20	
No. 1[8]	74.46	25.54	18.18	2.69	4.08	.59	
No. 2	64.80	35.20	25.40	7.61		2.19	
No. 3	56.50	43.50	31.57	8.44		3.49	
No. 4	61.67	38.33	33.43	2.62	1.56	.72	
Cream[9]	63.28	36.72	29.40	4.22	2.08	.40	.56

[1] Johnston's Ag. Chem. p. 548.
[2] Journ. R. A. S. 1863, pp. 317, 298.
[3] Lectures on Food, p 34.
[4] Quoted in Ag of O. 1858, p. 281.
[5] Sixth Rept. Am. Dairymen's Association, 1870, p. 160.
[6] Prof. Muller, quoted, Trans. Vt. Dairymen's Association, 1872, p. 150.
[7] Dr. Percy, Trans. Med. Soc. of State of N. Y. 1860, p. 47.
[8] Dr. Voelcker, Journ. R. A. S. xxiv, p 298.
[9] Dr. Hanneberg, quoted, Ag. of O., 1855, p. 282.

CREAM PERCENTAGE. 243

We find, from these analyses, that some creams may yield three times as much butter as other creams. In other words, that a milk yielding ten per cent of cream may furnish *more* butter than another milk indicating thirty per cent of cream. As the form in which the butter is held in the milk has much to do with the practical process of churning, and as it may be said that cream cannot vary to any very great extent in practice, it may be useful to quote the result obtained by Mr. Horsfall, in England, where a quart of cream yielded 16 ounces of butter at one time, and 22 to 24 ounces, and even 25 ounces at another. At the time that he obtained the largest result the indicated cream was but $6\frac{1}{2}$ per cent.

In order that our conclusions may be justified, we will offer some more figures. Prof. Caldwell[10] reports that Baumhauer, in Amsterdam, examined 20 different samples of milk in this manner. Nos. 1 and 3 were found by chemical analyses to have respectively 2.7 and 3.5 per cent of fat, while the cream-gauge indicated no difference between them. Nos. 5, 10, 15, 18, and 20 were found by the accurate chemical method to contain 3.3, 3.0, 3.9, 2.3, and 2.7 per cent of fat, but the thickness of the layer of cream formed by all of them was the same.

We have still another illustration derived from four cows' milk examined at different periods.[11]

---

[10] 7th Report Am. Dairymen's Association, 1871, p. 44.
[11] Prize Essay H. Soc. 1868–9, pp. 69, 70.

	April 23.		May 28.	
	Per cent Cream.	Per cent Butter.	Per cent Cream.	Per cent Butter.
No. 1	11.5	2.386	12.0	2.404
" 2	10.5	2.766	10.75	2.290
" 3	12.5	2.153	13.0	2.661
" 4	10.0	2.796	11.75	2.596

We are now prepared to assert that there is not necessarily any connection between the cream percentages and the butter yield. The holding forth of the large cream percentage yield of favored breeds or favored cows has no experimental relation whatever with their butter product. The modest cow, with a small cream percentage to her milk, *may* make more butter than the vaunted cow which is supposed to average 25 or 30 per cent of cream.

The whole system of claiming surpassing merit for a breed, through any one superficial feature, is an erroneous one. Of what practical use is a large cream percentage, if the relation of this statement to the butter product is not established?

A DIGRESSION.

Allow me to digress in order to call attention to a few errors. The first I shall take up is that the Jersey cow is a superior butter cow, *because* she yields a large percentage of cream. We have had no evidence offered for or against this point. Another error is in the statement that there is more of the chemical constituent caseine in the milk of the Ayrshire cow than in that from the Jersey cow. We have no evidence furnished to substantiate any predominance of caseine in Ayrshire milk. A grievous error for the dairymen is that of considering all cows alike which give milk, and either placing the native cow far ahead of thorough-breds for his use, or, rarely, the corresponding error of claiming that a thorough-bred, as a thorough-bred, is superior to the native. Now, when we consider that the thoroughbred and the native are of value only as they serve the uses of man, we must have a higher definition than simply purity of lineage. The thorough-bred derives its fame on account of being bred for a certain use, and hence is considered as a type for that use. We desire, therefore, to breed to perpetuate this type, which must be a useful one. The butter dairyman must seek the butter type of cow, and the cheese dairyman the cheese type, whether this seeking carries him towards the Ayrshire, the Jersey, the American Holstein, or the Short-horn. Let the dairyman discard prejudice as to the name of a cow, whether native or thorough-bred, and seek the substance in a type which is to be most useful to him. In doing this, those who breed their own calves will naturally make much use of the thorough-bred, and utility, not fancy, will settle the question as to which type or breed you shall seek. The importance of this digression consists in the *fact* that different milks have different qualities; and that in large herds, very often one or more particular cow's milk is adding but little towards the profit of the buttermaker, and might be withdrawn from the herd with a real advantage. The dairyman, if this be true, should seek a uniform type of cows. Is it not the want of uniformity in herds which can lead us to ask, Why is it that such good average results are as seldom obtained from the milk of a large herd as from a smaller one, except this matter of difference in the quality of milk, which in the larger herd has not been so readily perceived?

Coming back to our subject, we will again inquire,

WHAT IS CREAM?

It is the lighter portion of the milk, which is collected from the surface after standing. What more? It contains butter, some caseous matter, a little sugar of milk, some few salts, etc. We can give no proportional or more exact definition, on account of the great variations which may and do occur. This cream is affected differently by the souring changes which occur in it before it is placed in the churn. The cream from one class of cows may have its "churning time" hastened more by twenty-four hours' standing than another specimen of cream, from other cows, after having stood thirty-six hours, or even forty-eight. One cream will leave more waste in the buttermilk than will another cream. In other words, the analysis of the churn is not as complete, in every case, apart from the fat in the cream. One cream may churn "all in a lump." That is, the butter seems to "come" at about the same time throughout the whole mass of the cream. On the other hand, another cream will show specks of butter long before the general mass is churned. The explanation of these last two statements is the difference in the size of the globules. The more uniform their size, and the more uniform the strength or weakness of their membrane, the more accurately will they all rupture at one time, and allow the butter to collect. Again, as only the globules above a certain size are broken in the ordinary process of churning, the cream with the fewer granules (*i. e.* very small globules) would be expected to produce the most economical results.

CHURNING.

In the churn, cream also presents differences. One cream can be easily and quickly churned by a regular and even motion, while another may be benefited by, may even require, a more violent agitation. The dash churn, the barrel churn, the Blanchard or the Bullard, may each and all be the best churn possible under some circumstances. There is more difference between creams than between the best specimens of our churning machinery.

DEEP OR SHALLOW SETTING.

We now approach the disputed ground of deep or shallow setting of milk. Perhaps, pursuing our inquiries without prejudice and without prepossession, for we have experimented with neither, we may be able to determine the question theoretically, in a manner which may deserve the confidence of practice.

The form-element of the cream is the globule. This varies in size, and varies in specific gravity.

Being lighter than the fluid in which it occurs, the tendency of each globule is to seek the surface. The butter which we are striving to obtain is the pure fat of these globules, as free from foreign matter as may be, although in practice we find other substances in butter, as below: —

Butter.	Caseine.	Water.	Sugar.	Ash.	
84.75	.5025	13.695	.71	.095	Prof. Mueler.[12]
86.27	.94	12.79	..	..	Thompson.[13]
82.70	2.45	14.85	..	..	Prof. Way.[14]
76.67	3.38	16.95	..	..	do.[15]
79.12	3.37	17.51	..	..	do.[15]
94.4	.3	5.3	..	..	———.[16]
93.0	.3	6.7	..	..	do.[16]
87.5	1.0	11.5	..	..	do.[16]
78.5	.3	21.2	..	..	do.[16]

Let us see what would be the effect on mixing artificially, different sized bodies in water, and then, after well shaking, leaving the vessel containing the lot at rest. We will use sand for an illustration, because the principles being the same which underlie the process of acquiring an equilibrium, whether of lighter or heavier material, we have in this substance a handy one to study, and our conclusions can be readily verified.

EXPERIMENTS WITH SAND.

Suppose a handful of sand, of widely different sizes, to be violently shaken or stirred in a shallow, and also in a deep dish of water. What effect will

---

[12] Quoted in Trans. Vt. Dairymen's Association, 1872, p. 150.
[13] On Food of Animals, p. 63.
[14] Journ. R. A. S., xi. p. 735.
[15] Scalded Cream, Devonshire method, Journ. R. A. S. xi, p. 735.
[16] Wagner's Handbook of Chem. Tech. p. 559.

the depth of the dish have on the arrangement of the sand? It will be found that the particles of sand will arrange themselves according to their gravity or size as soon as the vessel comes to a rest. In the shallow dish the strata of sand thrown down will be somewhat mixed. If the other dish be sufficiently deep, the sand will be exactly graded and arranged with the heavier particles, or those which sink first, placed accurately at the bottom and the lighter particles at the top.

In deep setting of milk we have similar conditions, only reversed. The globules acted on by gravity arrange themselves in order, and the deeper the jar in which the milk is set, the more regularly will the globules arrange themselves according to size, from above down. In the shallow setting we have a greater mixture of the different sizes of globules in the same space than in the deeper setting, and the deeper the setting, the more completely would the globules be arranged in order according to size. Such being the physical effect, the question now arises, What effect will occur in practice from having cream of such different churning quality at different depths? for it will be remembered that the size of the globule has a strong influence on the churning.

Let us seek another illustration in the form of various sizes of shot. If we place a gill of coarse shot in a bowl, we can add a considerable quantity of fine shot without any increasing of bulk, as the fine shot will occupy the interstices between the larger pellets. So with the globules of the cream:

when there is considerable difference in their sizes as they accumulate together in rising, there must necessarily be more butter or fat, in a given bulk of cream, than when the globules are nearer of one size. As the globules are more completely mingled in the shallow setting than in the deeper, the cream of the shallow setting, bulk for bulk, should contain more butter than the cream from the same quality of milk that has had the use of a greater depth of fluid to arrange itself in.

In the arrangement of the globules in the deep setting, we have the larger globules and those which are easier churned at the top, and the smaller globules, which are churned with greater difficulty, at the bottom. That is, there is more difference between the upper and lower half of this cream than in the cream of the shallow set milk. In the cream, as placed in the churn, we have a more complete mixture of the various sized globules in the cream of the shallow setting than in the cream of the deep setting. With creams of the same elementary and physical composition, and of the same mechanical mixture, we should expect like results in an experimental trial. When, however, we see we have not the like mechanical mixture in these two methods of setting milk, we cannot expect equivalent results. Either the one or the other method must experimentally prove the better, according to the composition of the milk used. In order to illustrate the difficulty of obtaining cream or of dividing a sample of cream for the purpose of such experiments, let us refer to the writings of Mr. Horsfall,[17] who records that he took five quarts of cream in succession from a cream-pail and churned each batch separately.

  The first 5 quarts churned .... 127 ounces of butter.
  The second 5 quarts " .... 125 " "
  The third 5 quarts " .... 120½ " "

At a subsequent churning of 14 quarts of cream, the first 7 quarts gave 175 ounces of butter; the second 7, 177 ounces of butter.

Oftentimes, nay usually, when there is such a discrepancy of result between the two churnings from one pail of cream, but the operation carried on in churns of different makes, the difference which happens to be in favor of one churn is unhesitatingly ascribed to the superiority of the churn, and not to the superiority of the cream that the churn acts upon.

To carry on a series of experiments which should conclusively prove one method of setting milk preferable to another, in quantity and quality of butter, would consume much time and labor, and would be beyond the means of an ordinary dairyman. If, however, these principles as enunciated here are accurate, we have a foundation for a judgment which should be correct.

The known effect of gravity in arranging particles of differing sizes and weight, and the known influence of the globules of the milk in churning, lead me to assert, from a theoretical standpoint, that we

---

[17] Journ. R. A. S. 1856, p. 269.

should expect a larger proportion of butter from shallow setting than deep setting; but if there was any difference in quality, it would be in favor of the deep setting, provided there was neither over-setting nor over-churning.

The gist of my paper, and whatever importance it may deserve, is, that it attempts to show that cream is uniform neither in chemical, practical, nor physical composition; that, accordingly, the dairyman must treat each sample, or the average of that which comes under his care, through knowledge — that is, by science rather than routine; that dairy practices, in cases of difficulty, must be governed by reasoning rather than by guess-work. These remarks apply to milk as well as cream, — in fact, to the handling of all dairy products.

Imported "Model of Perfection."

Imported
"Pride of the Hills."

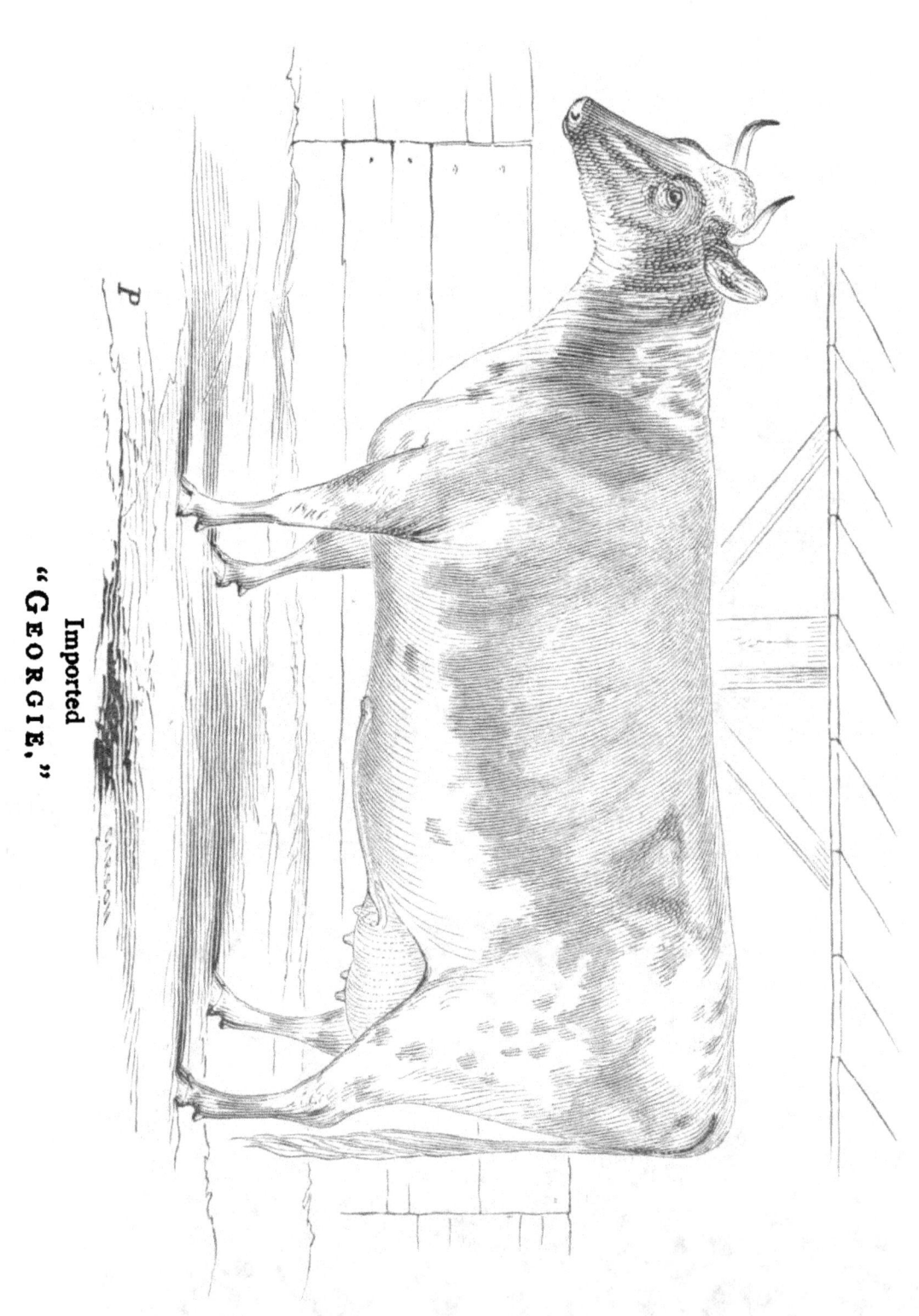

Imported "GEORGIE,"

www.ingramcontent.com/pod-product-compliance
Lightning Source LLC
Chambersburg PA
CBHW082322220526
45470CB00008B/2382